Antennas and Propagation: Technology and Applications

Antennas and Propagation: Technology and Applications

Edited by
Larry Graham

Larsen & Keller
www.larsen-keller.com

Antennas and Propagation: Technology and Applications
Edited by Larry Graham
ISBN 978-1-63549-028-2 (Hardback)

© 2017 Larsen & Keller

▤ Larsen & Keller

Published by Larsen and Keller Education,
5 Penn Plaza,
19th Floor,
New York, NY 10001, USA

Cataloging-in-Publication Data

Antennas and propagation : technology and applications / edited by Larry Graham.
    p. cm.
Includes bibliographical references and index.
ISBN 978-1-63549-028-2
1. Antennas (Electronics). 2. Radio--Antennas. 3. Radio wave propagation. I. Graham, Larry.
TK7871.6 .A58 2017
621.382 4--dc23

The publisher's policy is to use permanent paper from mills that operate a sustainable forestry policy. Furthermore, the publisher ensures that the text paper and cover boards used have met acceptable environmental accreditation standards.

Printed and bound in the United States of America.

For more information regarding Larsen and Keller Education and its products, please visit the publisher's website www.larsen-keller.com

Table of Contents

Preface

This book is a compilation of chapters that discuss the most vital concepts in the field of antennas and propagation. It talks about the uses and applications of this technology. An electrical device that converts radio waves into electric power and vice versa is known as an antenna or aerial. The devices which use this antenna technology are broadcast television, cell phones, communication receivers, radio broadcasting, wireless microphones, radars, bluetooth enabled devices and satellite communications, etc. This book presents the complex subject of antennas and propagation in the most comprehensible and easy to understand language. The topics included in it are of utmost significance and bound to provide incredible insights to readers. Those in search of information to further their knowledge will be greatly assisted by this text. It is a complete source of knowledge on the present status of this important field.

A foreword of all Chapters of the book is provided below:

Chapter 1 - The conversion of electric power into radio waves in an electrical device is known as an antenna whereas the change in modulation of radio waves from one point to another is denoted as radio propagation. They are used in systems such as broadcasting, radios, and cell phones as well as other devices such as wireless computers and bluetooth; **Chapter 2 -** The major components discussed in this chapter are transmitters, transposers, antenna tuners and radio receivers. A device used to rebroadcast signals in order for them to reach the receivers is known as a transposer while an antenna tuner to improves radio signals. To have a clear understanding, it's very important to comprehend the various devices of the antenna; **Chapter 3 -** A vital part of the digital wireless industry is the antenna. There are several different types of antennas; some of them which have been mentioned in the chapter such as monopole antenna, loop antenna, random wire antenna, smart antenna, etc.; **Chapter 4 -** Some of the theories stated are antenna farm, antenna aperture, antenna diversity and isotropic radiator. Antenna farm is the area dedicated to television and radio communications; usually an area with more than three antennas is referred to as an antenna farm. The chapter strategically encompasses and includes the essential components of antennas; **Chapter 5 -** When radio waves travel from one point to another, they exhibit characteristics determined by that location and such behavior is called propagation. Radio propagation can be affected by numerous influences determined by its path from A point to B point. This chapter explicates the key features of propagation by further explaining theories like shortwave radio, surface wave, two-way radio and skip zone.

I would like to thank the entire editorial team who made sincere efforts for this book and my family who supported me in my efforts of working on this book. I take this opportunity to thank all those who have been a guiding force throughout my life.

Editor

Introduction to Antenna and Propagation

The conversion of electric power into radio waves in an electrical device is known as an antenna whereas the change in modulation of radio waves from one point to another is denoted as radio propagation. They are used in systems such as broadcasting, radios, and cell phones as well as other devices such as wireless computers and bluetooth.

Antenna (Radio)

An antenna (plural antennae or antennas), or aerial, is an electrical device which converts electric power into radio waves, and vice versa. It is usually used with a radio transmitter or radio receiver. In transmission, a radio transmitter supplies an electric current oscillating at radio frequency (i.e. a high frequency alternating current (AC)) to the antenna's terminals, and the antenna radiates the energy from the current as electromagnetic waves (radio waves). In reception, an antenna intercepts some of the power of an electromagnetic wave in order to produce a tiny voltage at its terminals, that is applied to a receiver to be amplified.

Antennas are essential components of all equipment that uses radio. They are used in systems such as radio broadcasting, broadcast television, two-way radio, communications receivers, radar, cell phones, and satellite communications, as well as other devices such as garage door openers, wireless microphones, Bluetooth-enabled devices, wireless computer networks, baby monitors, and RFID tags on merchandise.

Typically an antenna consists of an arrangement of metallic conductors (elements), electrically connected (often through a transmission line) to the receiver or transmitter. An oscillating current of electrons forced through the antenna by a transmitter will create an oscillating magnetic field around the antenna elements, while the charge of the electrons also creates an oscillating electric field along the elements. These time-varying fields radiate away from the antenna into space as a moving transverse electromagnetic field wave. Conversely, during reception, the oscillating electric and magnetic fields of an incoming radio wave exert force on the electrons in the antenna elements, causing them to move back and forth, creating oscillating currents in the antenna.

Antennas can be designed to transmit and receive radio waves in all horizontal directions equally (omnidirectional antennas), or preferentially in a particular direction (directional or high gain antennas). In the latter case, an antenna may also include additional elements or surfaces with no electrical connection to the transmitter or receiver, such as parasitic elements, parabolic reflectors or horns, which serve to direct the radio waves into a beam or other desired radiation pattern.

The first antennas were built in 1888 by German physicist Heinrich Hertz in his pioneering experiments to prove the existence of electromagnetic waves predicted by the theory of James

Clerk Maxwell. Hertz placed dipole antennas at the focal point of parabolic reflectors for both transmitting and receiving. He published his work in *Annalen der Physik und Chemie* (vol. 36, 1889).

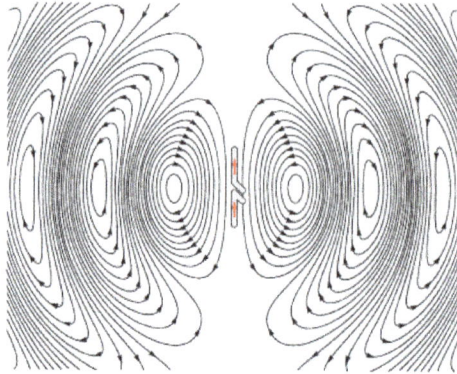

Animation of a half-wave dipole antenna transmitting radio waves, showing the electric field lines. The antenna in the center is two vertical metal rods, with an alternating current applied at its center from a radio transmitter *(not shown)*. The voltage charges the two sides of the antenna alternately positive *(+)* and negative *(−)*. Loops of electric field *(black lines)* leave the antenna and travel away at the speed of light; these are the radio waves.

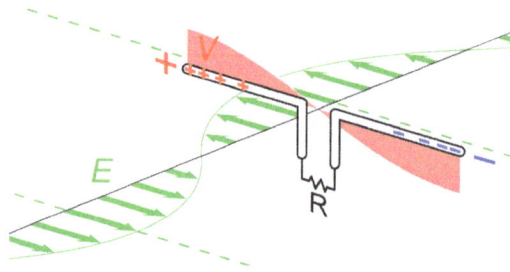

Animated diagram of a half-wave dipole antenna receiving energy from a radio wave. The antenna consists of two metal rods connected to a receiver *R*. The electric field *(E, green arrows)* of the incoming wave pushes the electrons in the rods back and forth, charging the ends alternately positive *(+)* and negative *(−)*. Since the length of the antenna is one half the wavelength of the wave, the oscillating field induces standing waves of voltage *(V, represented by red band)* and current in the rods. The oscillating currents *(black arrows)* flow down the transmission line and through the receiver (represented by the resistance *R*).

Terminology

The words *antenna* (plural: *antennas* in US English, although both "antennas" and "antennae" are used in International English) and *aerial* are used interchangeably. Occasionally the term "aerial" is used to mean a wire antenna. However, note the important international technical journal, the *IEEE Transactions on Antennas and Propagation*. In the United Kingdom and other areas where British English is used, the term aerial is sometimes used although 'antenna' has been universal in professional use for many years.

The origin of the word *antenna* relative to wireless apparatus is attributed to Italian radio pioneer Guglielmo Marconi. In the summer of 1895, Marconi began testing his wireless system outdoors on his father's estate near Bologna and soon began to experiment with long wire "aerials". Marconi discovered that by raising the "aerial" wire above the ground and connecting the other side of his transmitter to ground, the transmission range was increased. Soon he was able to transmit signals over a hill, a distance of approximately 2.4 kilometres (1.5 mi). In Italian a tent pole is known as *l'antenna centrale,* and the pole with the wire was simply called *l'antenna*. Until then wireless radiating transmitting and receiving elements were known simply as aerials or terminals.

Electronic symbol for an antenna

Because of his prominence, Marconi's use of the word *antenna* (Italian for *pole*) spread among wireless researchers, and later to the general public.

In common usage, the word *antenna* may refer broadly to an entire assembly including support structure, enclosure (if any), etc. in addition to the actual functional components. Especially at microwave frequencies, a receiving antenna may include not only the actual electrical antenna but an integrated preamplifier or mixer.

An antenna, in converting radio waves to electrical signals or vice versa, is a form of transducer.

Overview

Antennas of the Atacama Large Millimeter submillimeter Array.

Antennas are required by any radio receiver or transmitter to couple its electrical connection to the electromagnetic field. Radio waves are electromagnetic waves which carry signals through the air (or through space) at the speed of light with almost no transmission loss. Radio transmitters and receivers are used to convey signals (information) in systems including broadcast (audio) radio, television, mobile telephones, Wi-Fi (WLAN) data networks, trunk lines and point-to-point communications links (telephone, data networks), satellite links, many remote controlled devices such as garage door openers, and wireless remote sensors, among many others. Radio waves are also used directly for measurements in technologies including radar, GPS, and radio astronomy. In each and every case, the transmitters and receivers involved require antennas, although these are sometimes hidden (such as the antenna inside an AM radio or inside a laptop computer equipped with Wi-Fi).

According to their applications and technology available, antennas generally fall in one of two categories:

1. Omnidirectional or only weakly directional antennas which receive or radiate more or less in all directions. These are employed when the relative position of the other station is unknown or arbitrary. They are also used at lower frequencies where a directional antenna would be too large, or simply to cut costs in applications where a directional antenna isn't required.

2. Directional or *beam* antennas which are intended to preferentially radiate or receive in a particular direction or directional pattern.

In common usage "omnidirectional" usually refers to all horizontal directions, typically with reduced performance in the direction of the sky or the ground (a truly isotropic radiator is not even possible). A "directional" antenna usually is intended to maximize its coupling to the electromagnetic field in the direction of the other station, or sometimes to cover a particular sector such as a 120° horizontal fan pattern in the case of a panel antenna at a cell site.

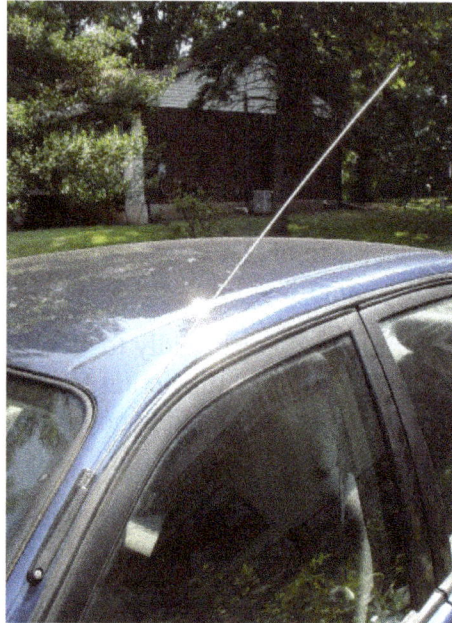

Whip antenna on car, common example of an omnidirectional antenna

One example of omnidirectional antennas is the very common *vertical antenna* or whip antenna consisting of a metal rod (often, but not always, a quarter of a wavelength long). A dipole antenna is similar but consists of two such conductors extending in opposite directions, with a total length that is often, but not always, a half of a wavelength long. Dipoles are typically oriented horizontally in which case they are weakly directional: signals are reasonably well radiated toward or received from all directions with the exception of the direction along the conductor itself; this region is called the antenna blind cone or null.

Both the vertical and dipole antennas are simple in construction and relatively inexpensive. The dipole antenna, which is the basis for most antenna designs, is a balanced component, with equal but opposite voltages and currents applied at its two terminals through a balanced transmission line (or to a coaxial transmission line through a so-called balun). The vertical antenna, on the other hand, is a *monopole* antenna. It is typically connected to the inner conductor of a coaxial transmission line (or a matching network); the shield of the transmission line is connected to ground. In this way, the ground (or any large conductive surface) plays the role of the second conductor of a dipole, thereby forming a complete circuit. Since monopole antennas rely on a conductive ground, a so-called grounding structure may be employed to provide a better ground contact to the earth or which itself acts as a ground plane to perform that function regardless of (or in absence of) an actual contact with the earth.

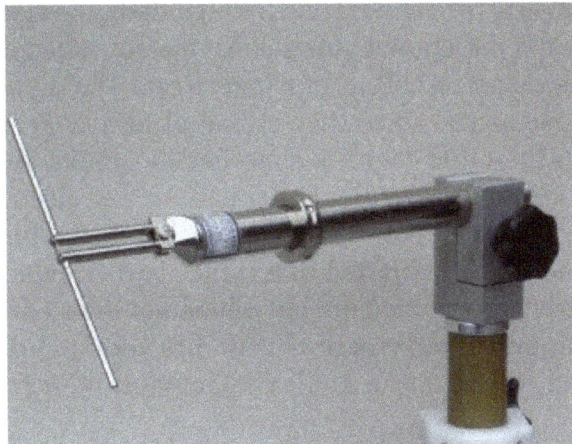

Half-wave dipole antenna

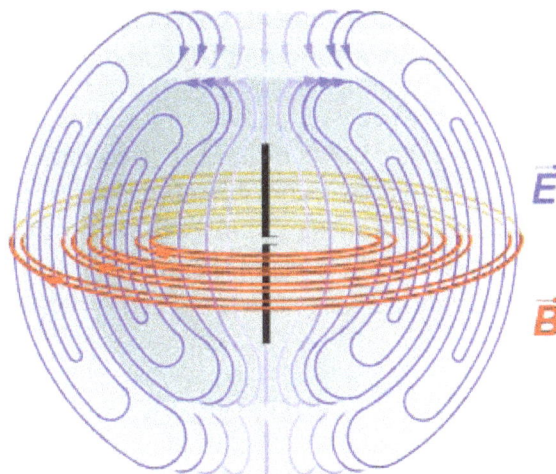

Diagram of the electric fields (blue) and magnetic fields (red) radiated by
a dipole antenna (black rods) during transmission.

Antennas more complex than the dipole or vertical designs are usually intended to increase the directivity and consequently the gain of the antenna. This can be accomplished in many different ways leading to a plethora of antenna designs. The vast majority of designs are fed with a balanced line (unlike a monopole antenna) and are based on the dipole antenna with additional components (or *elements*) which increase its directionality. Antenna "gain" in this instance describes the concentration of radiated power into a particular solid angle of space, as opposed to the spherically uniform radiation of the ideal radiator. The increased power in the desired direction is at the expense of that in the undesired directions. Power is conserved, and there is no net power increase over that delivered from the power source (the transmitter.)

For instance, a phased array consists of two or more simple antennas which are connected together through an electrical network. This often involves a number of parallel dipole antennas with a certain spacing. Depending on the relative phase introduced by the network, the same combination of dipole antennas can operate as a "broadside array" (directional normal to a line connecting the elements) or as an "end-fire array" (directional along the line connecting the elements). Antenna arrays may employ any basic (omnidirectional or weakly directional) antenna type, such as dipole, loop or slot antennas. These elements are often identical.

However a log-periodic dipole array consists of a number of dipole elements of *different* lengths in order to obtain a somewhat directional antenna having an extremely wide bandwidth: these are frequently used for television reception in fringe areas. The dipole antennas composing it are all considered "active elements" since they are all electrically connected together (and to the transmission line). On the other hand, a superficially similar dipole array, the Yagi-Uda Antenna (or simply "Yagi"), has only one dipole element with an electrical connection; the other so-called parasitic elements interact with the electromagnetic field in order to realize a fairly directional antenna but one which is limited to a rather narrow bandwidth. The Yagi antenna has similar looking parasitic dipole elements but which act differently due to their somewhat different lengths. There may be a number of so-called "directors" in front of the active element in the direction of propagation, and usually a single (but possibly more) "reflector" on the opposite side of the active element.

Greater directionality can be obtained using beam-forming techniques such as a parabolic reflector or a horn. Since high directivity in an antenna depends on it being large compared to the wavelength, narrow beams of this type are more easily achieved at UHF and microwave frequencies.

At low frequencies (such as AM broadcast), arrays of vertical towers are used to achieve directionality and they will occupy large areas of land. For reception, a long Beverage antenna can have significant directivity. For non directional portable use, a short vertical antenna or small loop antenna works well, with the main design challenge being that of impedance matching. With a vertical antenna a *loading coil* at the base of the antenna may be employed to cancel the reactive component of impedance; small loop antennas are tuned with parallel capacitors for this purpose.

An antenna lead-in is the transmission line (or *feed line*) which connects the antenna to a transmitter or receiver. The *antenna feed* may refer to all components connecting the antenna to the transmitter or receiver, such as an impedance matching network in addition to the transmission line. In a so-called aperture antenna, such as a horn or parabolic dish, the "feed" may also refer to a basic antenna inside the entire system (normally at the focus of the parabolic dish or at the throat of a horn) which could be considered the one active element in that antenna system. A microwave antenna may also be fed directly from a waveguide in place of a (conductive) transmission line.

An antenna counterpoise or ground plane is a structure of conductive material which improves or substitutes for the ground. It may be connected to or insulated from the natural ground. In a monopole antenna, this aids in the function of the natural ground, particularly where variations (or limitations) of the characteristics of the natural ground interfere with its proper function. Such a structure is normally connected to the return connection of an unbalanced transmission line such as the shield of a coaxial cable.

An electromagnetic wave *refractor* in some aperture antennas is a component which due to its shape and position functions to selectively delay or advance portions of the electromagnetic wavefront passing through it. The refractor alters the spatial characteristics of the wave on one side relative to the other side. It can, for instance, bring the wave to a focus or alter the wave front in other ways, generally in order to maximize the directivity of the antenna system. This is the radio equivalent of an optical lens.

Cell phone base station antennas

An antenna coupling network is a passive network (generally a combination of inductive and capacitive circuit elements) used for impedance matching in between the antenna and the transmitter or receiver. This may be used to improve the standing wave ratio in order to minimize losses in the transmission line and to present the transmitter or receiver with a standard resistive impedance that it expects to see for optimum operation.

Reciprocity

It is a fundamental property of antennas that the electrical characteristics of an antenna described in the next section, such as gain, radiation pattern, impedance, bandwidth, resonant frequency and polarization, are the same whether the antenna is transmitting or receiving. For example, the "*receiving pattern*" (sensitivity as a function of direction) of an antenna when used for reception is identical to the radiation pattern of the antenna when it is *driven* and functions as a radiator. This is a consequence of the reciprocity theorem of electromagnetics. Therefore, in discussions of antenna properties no distinction is usually made between receiving and transmitting terminology, and the antenna can be viewed as either transmitting or receiving, whichever is more convenient.

A necessary condition for the aforementioned reciprocity property is that the materials in the antenna and transmission medium are linear and reciprocal. *Reciprocal* (or *bilateral*) means that the material has the same response to an electric current or magnetic field in one direction, as it has to the field or current in the opposite direction. Most materials used in antennas meet these conditions, but some microwave antennas use high-tech components such as isolators and circulators, made of nonreciprocal materials such as ferrite. These can be used to give the antenna a different behavior on receiving than it has on transmitting, which can be useful in applications like radar.

Characteristics

Antennas are characterized by a number of performance measures which a user would be concerned with in selecting or designing an antenna for a particular application. Chief among these relate to the directional characteristics (as depicted in the antenna's *radiation pattern*) and the resulting *gain*. Even in omnidirectional (or weakly directional) antennas, the gain can often be increased by concentrating more of its power in the horizontal directions, sacrificing power radiated toward the sky and ground. The antenna's power gain (or simply "gain") also takes into account the antenna's efficiency, and is often the primary figure of merit.

Resonant antennas are expected to be used around a particular *resonant frequency*; an antenna must therefore be built or ordered to match the frequency range of the intended application. A particular antenna design will present a particular feedpoint impedance. While this may affect the choice of an antenna, an antenna's impedance can also be adapted to the desired impedance level of a system using a matching network while maintaining the other characteristics (except for a possible loss of efficiency).

Although these parameters can be measured in principle, such measurements are difficult and require very specialized equipment. Beyond tuning a transmitting antenna using an SWR meter, the typical user will depend on theoretical predictions based on the antenna design or on claims of a vendor.

An antenna transmits and receives radio waves with a particular polarization which can be reoriented by tilting the axis of the antenna in many (but not all) cases. The physical size of an antenna is often a practical issue, particularly at lower frequencies (longer wavelengths). Highly directional antennas need to be significantly larger than the wavelength. Resonant antennas usually use a linear conductor (or *element*), or pair of such elements, each of which is about a quarter of the wavelength in length (an odd multiple of quarter wavelengths will also be resonant). Antennas that are required to be small compared to the wavelength sacrifice efficiency and cannot be very directional. Fortunately at higher frequencies (UHF, microwaves) trading off performance to obtain a smaller physical size is usually not required.

Resonant Antennas

The majority of antenna designs are based on the *resonance* principle. This relies on the behaviour of moving electrons, which reflect off surfaces where the dielectric constant changes, in a fashion similar to the way light reflects when optical properties change. In these designs, the reflective surface is created by the end of a conductor, normally a thin metal wire or rod, which in the simplest case has a *feed point* at one end where it is connected to a transmission line. The conductor, or *element*, is aligned with the electrical field of the desired signal, normally meaning it is perpendicular to the line from the antenna to the source (or receiver in the case of a broadcast antenna).

The radio signal's electrical component induces a voltage in the conductor. This causes an electrical current to begin flowing in the direction of the signal's instantaneous field. When the resulting current reaches the end of the conductor, it reflects, which is equivalent to a 180 degree change in phase. If the conductor is $\frac{1}{4}$ of a wavelength long, current from the feed point will undergo 90 degree phase change by the time it reaches the end of the conductor, reflect through 180 degrees, and then another 90 degrees as it travels back. That means it has undergone a total 360 degree phase change, returning it to the original signal. The current in the element thus adds to the current being created from the source at that instant. This process creates a standing wave in the conductor, with the maximum current at the feed.

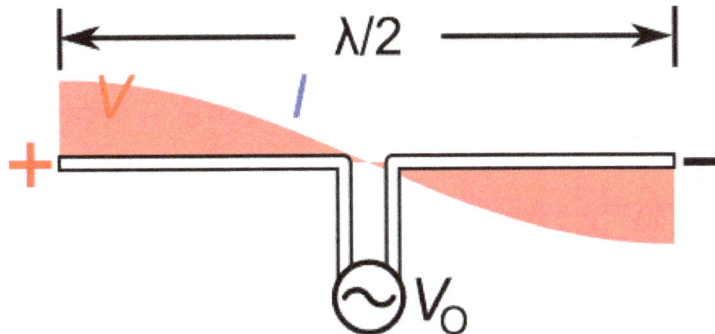

Standing waves on a half wave dipole driven at its resonant frequency. The waves are shown graphically by bars of color (red for voltage, V and blue for current, I) whose width is proportional to the amplitude of the quantity at that point on the antenna.

The ordinary half-wave dipole is probably the most widely used antenna design. This consists of two $\frac{1}{4}$-wavelength elements arranged end-to-end, and lying along essentially the same axis (or *collinear*), each feeding one side of a two-conductor transmission wire. The physical arrangement of the two elements places them 180 degrees out of phase, which means that at any given instant one of the elements is driving current into the transmission line while the other is pulling it out. The monopole antenna is essentially one half of the half-wave dipole, a single $\frac{1}{4}$-wavelength element with the other side connected to ground or an equivalent ground plane (or *counterpoise*). Monopoles, which are one-half the size of a dipole, are common for long-wavelength radio signals where a dipole would be impractically large. Another common design is the folded dipole, which is essentially two dipoles placed side-by-side and connected at their ends to make a single one-wavelength antenna.

The standing wave forms with this desired pattern at the design frequency, f_o, and antennas are normally designed to be this size. However, feeding that element with $3f_o$ (whose wavelength is $\frac{1}{3}$ that of f_o) will also lead to a standing wave pattern. Thus, an antenna element is *also* resonant when its length is $\frac{3}{4}$ of a wavelength. This is true for all odd multiples of $\frac{1}{4}$ wavelength. This allows some flexibility of design in terms of antenna lengths and feed points. Antennas used in such a fashion are known to be *harmonically operated*.

Current and Voltage Distribution

The quarter-wave elements imitate a series-resonant electrical element due to the standing wave present along the conductor. At the resonant frequency, the standing wave has a current peak and voltage node (minimum) at the feed. In electrical terms, this means the element has minimum reactance, generating the maximum current for minimum voltage. This is the ideal situation, because it produces the maximum output for the minimum input, producing the highest possible efficiency. Contrary to an ideal (lossless) series-resonant circuit, a finite resistance remains (corresponding to the relatively small voltage at the feed-point) due to the antenna's radiation resistance as well as any actual electrical losses.

Recall that a current will reflect when there are changes in the electrical properties of the material.

In order to efficiently send the signal into the transmission line, it is important that the transmission line has the same impedance as the elements, otherwise some of the signal will be reflected back into the antenna. This leads to the concept of impedance matching, the design of the overall system of antenna and transmission line so the impedance is as close as possible, thereby reducing these losses. Impedance matching between antennas and transmission lines is commonly handled through the use of a balun, although other solutions are also used in certain roles. An important measure of this basic concept is the standing wave ratio, which measures the magnitude of the reflected signal.

Consider a half-wave dipole designed to work with signals 1 m wavelength, meaning the antenna would be approximately 50 cm across. If the element has a length-to-diameter ratio of 1000, it will have an inherent resistance of about 63 ohms. Using the appropriate transmission wire or balun, we match that resistance to ensure minimum signal loss. Feeding that antenna with a current of 1 ampere will require 63 volts of RF, and the antenna will radiate 63 watts (ignoring losses) of radio frequency power. Now consider the case when the antenna is fed a signal with a wavelength of 1.25 m; in this case the reflected current would arrive at the feed out-of-phase with the signal, causing the net current to drop while the voltage remains the same. Electrically this appears to be a very high impedance. The antenna and transmission line no longer have the same impedance, and the signal will be reflected back into the antenna, reducing output. This could be addressed by changing the matching system between the antenna and transmission line, but that solution only works well at the new design frequency.

The end result is that the resonant antenna will efficiently feed a signal into the transmission line only when the source signal's frequency is close to that of the design frequency of the antenna, or one of the resonant multiples. This makes resonant antenna designs inherently narrowband, and they are most commonly used with a single target signal. They are particularly common on radar systems, where the same antenna is used for both broadcast and reception, or for radio and television *broadcasts*, where the antenna is working with a single frequency. They are less commonly used for reception where multiple channels are present, in which case additional modifications are used to increase the bandwidth, or entirely different antenna designs are used.

Modified Resonant Designs

It is possible to use simple impedance matching concepts to allow the use of monopole or dipole antennas substantially shorter than the ¼ or ½ wavelength, respectively, at which they are resonant. As these antennas are made shorter (for a given frequency) their impedance becomes dominated by a series capacitive (negative) reactance; by adding a series inductance with the opposite (positive) reactance – a so-called loading coil – the antenna's reactance may be cancelled leaving only a pure resistance. Sometimes the resulting (lower) electrical resonant frequency of such a system (antenna plus matching network) is described using the construct of *electrical length*, so the use of an antenna at a lower frequency than its resonant frequency may be termed *electrical lengthening*.

For example, at 30 MHz (10 m wavelength) a true resonant ¼ wavelength monopole would be almost 2.5 meters long, and using an antenna only 1.5 meters tall would require the addition of a loading coil. Then it may be said that the coil has lengthened the antenna to achieve an electrical length of 2.5 meters. However, the resulting resistive impedance achieved will be quite a bit lower than that of a true ¼ wave (resonant) monopole, often requiring further impedance matching (a transformer) to the desired transmission line. For ever shorter antennas (requiring greater "electrical lengthening") the radiation resistance plummets (approximately according to the square of the antenna length), so that the mismatch due to a net reactance away from the electrical resonance worsens. Or one could as well say that the equivalent resonant circuit of the antenna system has a higher Q factor and thus a reduced bandwidth, which can even become inadequate for the transmitted signal's spectrum. Resistive losses due to the loading coil, relative to the decreased radiation resistance, entail a reduced electrical efficiency, which can be of great concern for a transmitting antenna, but bandwidth is the major factor that sets the size of antennas at 1 MHz and lower frequencies.

Arrays and Reflectors

Rooftop television Yagi-Uda antennas like these are widely used at VHF and UHF frequencies.

The amount of signal received from a distant transmission source is essentially geometric in nature due to the inverse-square law, and this leads to the concept of *effective area*. This measures the performance of an antenna by comparing the amount of power it generates to the amount of power in the original signal, measured in terms of the signal's power density in Watts per square metre. A half-wave dipole has an effective area of 0.13 ^{2}. If more performance is needed, one cannot simply make the antenna larger. Although this would intercept more energy from the signal, due to the considerations above, it would decrease the output significantly due to it moving away from the resonant length. In roles where higher performance is needed, designers often use multiple elements combined together.

Returning to the basic concept of current flows in a conductor, consider what happens if a half-wave dipole is not connected to a feed point, but instead shorted out. Electrically this forms a single $\frac{1}{2}$-wavelength element. But the overall current pattern is the same; the current will be zero at the two ends, and reach a maximum in the center. Thus signals near the design frequency will continue to create a standing wave pattern. Any varying electrical current, like the standing wave in the element, will radiate a signal. In this case, aside from resistive losses in the element, the rebroadcast signal will be significantly similar to the original signal in both magnitude and shape. If this element is placed so its signal reaches the main dipole in-phase, it will reinforce the original signal, and increase the current in the dipole. Elements used in this way are known as *passive elements*.

A Yagi-Uda array uses passive elements to greatly increase gain. It is built along a support boom that is pointed toward the signal, and thus sees no induced signal and does not contribute to the antenna's operation. The end closer to the source is referred to as the front. Near the rear is a single active element, typically a half-wave dipole or folded dipole. Passive elements are arranged in front (*directors*) and behind (*reflectors*) the active element along the boom. The Yagi has the inherent quality that it becomes increasingly directional, and thus has higher gain, as the number of elements increases. However, this also makes it increasingly sensitive to changes in frequency; if the signal frequency changes, not only does the active element receive less energy directly, but all of the passive elements adding to that signal also decrease their output as well and their signals no longer reach the active element in-phase.

It is also possible to use multiple active elements and combine them together with transmission lines to produce a similar system where the phases add up to reinforce the output. The antenna array and very similar reflective array antenna consist of multiple elements, often half-wave dipoles, spaced out on a plane and wired together with transmission lines with specific phase lengths to produce a single in-phase signal at the output. The log-periodic antenna is a more complex design that uses multiple in-line elements similar in appearance to the Yagi-Uda but using transmission lines between the elements to produce the output.

Reflection of the original signal also occurs when it hits an extended conductive surface, in a fashion similar to a mirror. This effect can also be used to increase signal through the use of a *reflector*, normally placed behind the active element and spaced so the reflected signal reaches the element in-phase. Generally the reflector will remain highly reflective even if it is not solid; gaps less than $\frac{1}{10}$ generally have little effect on the outcome. For this reason, reflectors often take the form of wire meshes or rows of passive elements, which makes them lighter and less subject to wind. The parabolic reflector is perhaps the best known example of a reflector-based antenna, which has an effective area far greater than the active element alone.

Bandwidth

Although a resonant antenna has a purely resistive feed-point impedance at a particular frequency, many (if not most) applications require using an antenna over a range of frequencies. The frequency range or *bandwidth* over which an antenna functions well can be very wide (as in a log-periodic antenna) or narrow (in a resonant antenna); outside this range the antenna impedance becomes a poor match to the transmission line and transmitter (or receiver). Also in the case of the Yagi-Uda and other end-fire arrays, use of the antenna well away from its design frequency affects its radiation pattern, reducing its directive gain; the usable bandwidth is then limited regardless of impedance matching.

Except for the latter concern, the resonant frequency of an antenna system can always be altered by adjusting a suitable matching network. This is most efficiently accomplished using a matching network at the site of the antenna, since simply adjusting a matching network at the transmitter (or receiver) would leave the transmission line with a poor standing wave ratio.

Instead, it is often desired to have an antenna whose impedance does not vary so greatly over a certain bandwidth. It turns out that the amount of reactance seen at the terminals of a resonant antenna when the frequency is shifted, say, by 5%, depends very much on the diameter of the conductor used. A long thin wire used as a half-wave dipole (or quarter wave monopole) will have a reactance significantly greater than the resistive impedance it has at resonance, leading to a poor match and generally unacceptable performance. Making the element using a tube of a diameter perhaps 1/50 of its length, however, results in a reactance at this altered frequency which is not so great, and a much less serious mismatch and effect on the antenna's net performance. Thus rather thick tubes are often used for the elements; these also have reduced parasitic resistance (loss).

Rather than just using a thick tube, there are similar techniques used to the same effect such as replacing thin wire elements with *cages* to simulate a thicker element. This widens the bandwidth of the resonance. On the other hand, it is desired for amateur radio antennas to operate at several bands which are widely separated from each other (but not in between). This can often be accomplished simply by connecting elements resonant at those different frequencies in parallel. Most of the transmitter's power will flow into the resonant element while the others present a high (reactive) impedance, thus drawing little current from the same voltage. Another popular solution uses so-called *traps* consisting of parallel resonant circuits which are strategically placed in breaks along each antenna element. When used at one particular frequency band the trap presents a very high impedance (parallel resonance) effectively truncating the element at that length, making it a proper resonant antenna. At a lower frequency the trap allows the full length of the element to be employed, albeit with a shifted resonant frequency due to the inclusion of the trap's net reactance at that lower frequency.

The bandwidth characteristics of a resonant antenna element can be characterized according to its Q, just as one uses to characterize the sharpness of an L-C resonant circuit. A common mistake is to assume that there is an advantage in an antenna having a high Q (the so-called "quality factor"). In the context of electronic circuitry a low Q generally signifies greater loss (due to unwanted resistance) in a resonant L-C circuit, and poorer receiver selectivity. However this understanding does not apply to resonant antennas where the resistance involved is the radiation resistance, a desired quantity which removes energy from the resonant element in order to radiate it (the purpose of an antenna, after all!). The Q of an L-C-R circuit is defined as the ratio of the inductor's (or capacitor's) reactance to the resistance, so for a certain radiation resistance (the radiation resistance at resonance does not vary greatly with diameter) the greater reactance off-resonance causes the poorer bandwidth of an antenna employing a very thin conductor. The Q of such a narrowband antenna can be as high as 15. On the other hand, the reactance at the same off-resonant frequency of one using thick elements is much less, consequently resulting in a Q as low as 5. These two antennas may perform equivalently at the resonant frequency, but the second antenna will perform over a bandwidth 3 times as wide as the antenna consisting of a thin conductor.

Antennas for use over much broader frequency ranges are achieved using further techniques. Adjustment of a matching network can, in principle, allow for any antenna to be matched at any frequency. Thus the loop antenna built into most AM broadcast (medium wave) receivers has a very narrow bandwidth, but is tuned using a parallel capacitance which is adjusted according to the receiver tuning. On the other hand, log-periodic antennas are *not* resonant at any frequency but can be built to attain similar characteristics (including feedpoint impedance) over any frequency range. These are therefore commonly used (in the form of directional log-periodic dipole arrays) as television antennas.

Gain

Gain is a parameter which measures the degree of directivity of the antenna's radiation pattern. A high-gain antenna will radiate most of its power in a particular direction, while a low-gain antenna will radiate over a wider angle. The *antenna gain,* or *power gain* of an antenna is defined as the ratio of the intensity (power per unit surface area) I radiated by the antenna in the direction of its maximum output, at an arbitrary distance, divided by the intensity I_{iso} radiated at the same distance by a hypothetical isotropic antenna which radiates equal power in all directions. This dimensionless ratio is usually expressed logarithmically in decibels, these units are called "decibels-isotropic" (dBi)

$$G_{dBi} = 10\log\frac{I}{I_{iso}}$$

A second unit used to measure gain is the ratio of the power radiated by the antenna to the power radiated by a half-wave dipole antenna I_{dipole}; these units are called "decibels-dipole" (dBd)

$$G_{dBd} = 10\log\frac{I}{I_{dipole}}$$

Since the gain of a half-wave dipole is 2.15 dBi and the logarithm of a product is additive, the gain in dBi is just 2.15 decibels greater than the gain in dBd

$$G_{dBi} = G_{dBd} + 2.15$$

High-gain antennas have the advantage of longer range and better signal quality, but must be aimed carefully at the other antenna. An example of a high-gain antenna is a parabolic dish such as a satellite television antenna. Low-gain antennas have shorter range, but the orientation of the antenna is relatively unimportant. An example of a low-gain antenna is the whip antenna found on portable radios and cordless phones. Antenna gain should not be confused with amplifier gain, a separate parameter measuring the increase in signal power due to an amplifying device.

Effective Area or Aperture

The *effective area* or effective aperture of a receiving antenna expresses the portion of the power of a passing electromagnetic wave which it delivers to its terminals, expressed in terms of an equivalent area. For instance, if a radio wave passing a given location has a flux of 1 pW / m² (10^{-12} watts per square meter) and an antenna has an effective area of 12 m², then the antenna would deliver 12 pW of RF power to the receiver (30 microvolts rms at 75 ohms). Since the receiving antenna is not equally sensitive to signals received from all directions, the effective area is a function of the direction to the source.

Due to reciprocity (discussed above) the gain of an antenna used for transmitting must be proportional to its effective area when used for receiving. Consider an antenna with no loss, that is, one whose electrical efficiency is 100%. It can be shown that its effective area averaged over all directions must be equal to $\lambda^2/4\pi$, the wavelength squared divided by 4π. Gain is defined such that the average gain over all directions for an antenna with 100% electrical efficiency is equal to 1.

Therefore, the effective area A_{eff} in terms of the gain G in a given direction is given by:

$$A_{\text{eff}} = \frac{\lambda^2}{4\pi} G$$

For an antenna with an efficiency of less than 100%, both the effective area and gain are reduced by that same amount. Therefore, the above relationship between gain and effective area still holds. These are thus two different ways of expressing the same quantity. A_{eff} is especially convenient when computing the power that would be received by an antenna of a specified gain, as illustrated by the above example.

Radiation Pattern

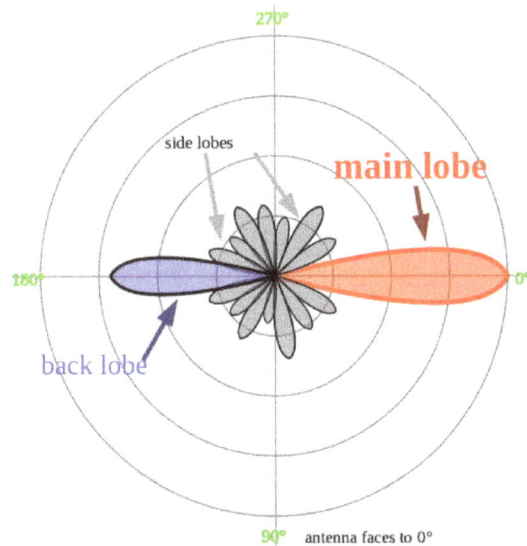

Polar plots of the horizontal cross sections of a (virtual) Yagi-Uda-antenna.Outline connects points with 3db field power compared to an ISO emitter.

The radiation pattern of an antenna is a plot of the relative field strength of the radio waves emitted by the antenna at different angles. It is typically represented by a three-dimensional graph, or polar plots of the horizontal and vertical cross sections. The pattern of an ideal isotropic antenna, which radiates equally in all directions, would look like a sphere. Many nondirectional antennas, such as monopoles and dipoles, emit equal power in all horizontal directions, with the power dropping off at higher and lower angles; this is called an omnidirectional pattern and when plotted looks like a torus or donut.

The radiation of many antennas shows a pattern of maxima or "*lobes*" at various angles, separated by "*nulls*", angles where the radiation falls to zero. This is because the radio waves emitted by different parts of the antenna typically interfere, causing maxima at angles where the radio waves arrive at distant points in phase, and zero radiation at other angles where the radio waves arrive out of phase. In a directional antenna designed to project radio waves in a particular direction, the lobe in that direction is designed larger than the others and is called the "*main lobe*". The other lobes usually represent unwanted radiation and are called "*sidelobes*". The axis through the main lobe is called the "*principal axis*" or "*boresight axis*".

Field Regions

The space surrounding an antenna can be divided into three concentric regions: the reactive near-field, the radiating near-field (Fresnel region) and the far-field (Fraunhofer) regions. These regions are useful to identify the field structure in each, although there are no precise boundaries.

In the far-field region, we are far enough from the antenna to neglect its size and shape. We can assume that the electromagnetic wave is purely a radiating plane wave (electric and magnetic fields are in phase and perpendicular to each other and to the direction of propagation). This simplifies the mathematical analysis of the radiated field.

Impedance

As an electro-magnetic wave travels through the different parts of the antenna system (radio, feed line, antenna, free space) it may encounter differences in impedance (E/H, V/I, etc.). At each interface, depending on the impedance match, some fraction of the wave's energy will reflect back to the source, forming a standing wave in the feed line. The ratio of maximum power to minimum power in the wave can be measured and is called the standing wave ratio (SWR). A SWR of 1:1 is ideal. A SWR of 1.5:1 is considered to be marginally acceptable in low power applications where power loss is more critical, although an SWR as high as 6:1 may still be usable with the right equipment. Minimizing impedance differences at each interface (impedance matching) will reduce SWR and maximize power transfer through each part of the antenna system.

Complex impedance of an antenna is related to the electrical length of the antenna at the wavelength in use. The impedance of an antenna can be matched to the feed line and radio by adjusting the impedance of the feed line, using the feed line as an impedance transformer. More commonly, the impedance is adjusted at the load with an antenna tuner, a balun, a matching transformer, matching networks composed of inductors and capacitors, or matching sections such as the gamma match.

Efficiency

Efficiency of a transmitting antenna is the ratio of power actually radiated (in all directions) to the power absorbed by the antenna terminals. The power supplied to the antenna terminals which is not radiated is converted into heat. This is usually through loss resistance in the antenna's conductors, but can also be due to dielectric or magnetic core losses in antennas (or antenna systems) using such components. Such loss effectively robs power from the transmitter, requiring a stronger transmitter in order to transmit a signal of a given strength.

For instance, if a transmitter delivers 100 W into an antenna having an efficiency of 80%, then the antenna will radiate 80 W as radio waves and produce 20 W of heat. In order to radiate 100 W of power, one would need to use a transmitter capable of supplying 125 W to the antenna. Note that antenna efficiency is a separate issue from impedance matching, which may also reduce the amount of power radiated using a given transmitter. If an SWR meter reads 150 W of incident power and 50 W of reflected power, that means that 100 W have actually been absorbed by the antenna (ignoring transmission line losses). How much of that power has actually been radiated cannot be directly determined through electrical measurements at (or before) the antenna terminals, but would require (for instance) careful measurement of field strength. Fortunately the loss resistance of antenna conductors such as aluminum rods can be calculated and the efficiency of an antenna using such materials predicted.

However loss resistance will generally affect the feedpoint impedance, adding to its resistive (real) component. That resistance will consist of the sum of the radiation resistance R_r and the loss resistance R_{loss}. If an rms current I is delivered to the terminals of an antenna, then a power of I^2R_r will be radiated and a power of I^2R_{loss} will be lost as heat. Therefore, the efficiency of an antenna is equal to $R_r / (R_r + R_{loss})$. Of course only the total resistance $R_r + R_{loss}$ can be directly measured.

According to reciprocity, the efficiency of an antenna used as a receiving antenna is identical to the efficiency as defined above. The power that an antenna will deliver to a receiver (with a proper impedance match) is reduced by the same amount. In some receiving applications, the very inefficient antennas may have little impact on performance. At low frequencies, for example, atmospheric or man-made noise can mask antenna inefficiency. For example, CCIR Rep. 258-3 indicates man-made noise in a residential setting at 40 MHz is about 28 dB above the thermal noise floor. Consequently, an antenna with a 20 dB loss (due to inefficiency) would have little impact on system noise performance. The loss within the antenna will affect the intended signal and the noise/interference identically, leading to no reduction in signal to noise ratio (SNR).

This is fortunate, since antennas at lower frequencies which are not rather large (a good fraction of a wavelength in size) are inevitably inefficient (due to the small radiation resistance R_r of small antennas). Most AM broadcast radios (except for car radios) take advantage of this principle by including a small loop antenna for reception which has an extremely poor efficiency. Using such an inefficient antenna at this low frequency (530–1650 kHz) thus has little effect on the receiver's net performance, but simply requires greater amplification by the receiver's electronics. Contrast this tiny component to the massive and very tall towers used at AM broadcast stations for transmitting at the very same frequency, where every percentage point of reduced antenna efficiency entails a substantial cost.

The definition of antenna gain or *power gain* already includes the effect of the antenna's efficiency. Therefore, if one is trying to radiate a signal toward a receiver using a transmitter of a given power, one need only compare the gain of various antennas rather than considering the efficiency as well. This is likewise true for a receiving antenna at very high (especially microwave) frequencies, where the point is to receive a signal which is strong compared to the receiver's noise temperature. However, in the case of a directional antenna used for receiving signals with the intention of *rejecting* interference from different directions, one is no longer concerned with the antenna efficiency, as discussed above. In this case, rather than quoting the antenna gain, one would be more concerned with the *directive gain* which does *not* include the effect of antenna (in) efficiency. The directive gain of an antenna can be computed from the published gain divided by the antenna's efficiency.

Polarization

The *polarization* of an antenna refers to the orientation of the electric field (E-plane) of the radio wave with respect to the Earth's surface and is determined by the physical structure of the antenna and by its orientation; note that this designation is totally distinct from the antenna's directionality. Thus, a simple straight wire antenna will have one polarization when mounted vertically, and a different polarization when mounted horizontally. As a transverse wave, the magnetic field of a radio wave is at right angles to that of the electric field, but by convention, talk of an antenna's "polarization" is understood to refer to the direction of the electric field.

Reflections generally affect polarization. For radio waves, one important reflector is the ionosphere which can change the wave's polarization. Thus for signals received following reflection by the ionosphere (a skywave), a consistent polarization cannot be expected. For line-of-sight communications or ground wave propagation, horizontally or vertically polarized transmissions generally remain in about the same polarization state at the receiving location. Matching the receiving antenna's polarization to that of the transmitter can make a very substantial difference in received signal strength.

Polarization is predictable from an antenna's geometry, although in some cases it is not at all obvious (such as for the quad antenna). An antenna's linear polarization is generally along the direction (as viewed from the receiving location) of the antenna's currents when such a direction can be defined. For instance, a vertical whip antenna or Wi-Fi antenna vertically oriented will transmit and receive in the vertical polarization. Antennas with horizontal elements, such as most rooftop TV antennas in the United States, are horizontally polarized (broadcast TV in the U.S. usually uses horizontal polarization). Even when the antenna system has a vertical orientation, such as an array of horizontal dipole antennas, the polarization is in the horizontal direction corresponding to the current flow. The polarization of a commercial antenna is an essential specification.

Polarization is the sum of the E-plane orientations over time projected onto an imaginary plane perpendicular to the direction of motion of the radio wave. In the most general case, polarization is elliptical, meaning that the polarization of the radio waves varies over time. Two special cases are linear polarization (the ellipse collapses into a line) as we have discussed above, and circular polarization (in which the two axes of the ellipse are equal). In linear polarization the electric field of the radio wave oscillates back and forth along one direction; this can be affected by the mounting of the antenna but usually the desired direction is either horizontal or vertical polarization. In circular polarization, the electric field (and magnetic field) of the radio wave rotates at the radio frequency circularly around the axis of propagation. Circular or elliptically polarized radio waves are designated as right-handed or left-handed using the "thumb in the direction of the propagation" rule. Note that for circular polarization, optical researchers use the opposite right hand rule from the one used by radio engineers.

It is best for the receiving antenna to match the polarization of the transmitted wave for optimum reception. Intermediate matchings will lose some signal strength, but not as much as a complete mismatch. A circularly polarized antenna can be used to equally well match vertical or horizontal linear polarizations. Transmission from a circularly polarized antenna received by a linearly polarized antenna (or vice versa) entails a 3 dB reduction in signal-to-noise ratio as the received power has thereby been cut in half.

Impedance Matching

Maximum power transfer requires matching the impedance of an antenna system (as seen looking into the transmission line) to the complex conjugate of the impedance of the receiver or transmitter. In the case of a transmitter, however, the desired matching impedance might not correspond to the dynamic output impedance of the transmitter as analyzed as a source impedance but rather the design value (typically 50 ohms) required for efficient and safe operation of the transmitting circuitry. The intended impedance is normally resistive but a transmitter (and some receivers) may have additional adjustments to cancel a certain amount of reactance in order to "tweak" the match. When a transmission line is used in between the antenna and the transmitter (or receiver) one generally would like an antenna system whose impedance is resistive and near the characteristic impedance of that transmission line in order to minimize the standing wave ratio (SWR) and the increase in transmission line losses it entails, in addition to supplying a good match at the transmitter or receiver itself.

Antenna tuning generally refers to cancellation of any reactance seen at the antenna terminals, leaving only a resistive impedance which might or might not be exactly the desired impedance (that of the transmission line). Although an antenna may be designed to have a purely resistive feedpoint impedance (such as a dipole 97% of a half wavelength long) this might not be exactly true at the frequency that it is eventually used at. In some cases the physical length of the antenna can be "trimmed" to obtain a pure resistance. On the other hand, the addition of a series inductance or parallel capacitance can be used to cancel a residual capacitive or inductive reactance, respectively.

In some cases this is done in a more extreme manner, not simply to cancel a small amount of residual reactance, but to resonate an antenna whose resonance frequency is quite different from the intended frequency of operation. For instance, a "whip antenna" can be made significantly shorter than 1/4 wavelength long, for practical reasons, and then resonated using a so-called loading coil. This physically large inductor at the base of the antenna has an inductive reactance which is the opposite of the capacitative reactance that such a vertical antenna has at the desired operating frequency. The result is a pure resistance seen at feedpoint of the loading coil; unfortunately that resistance is somewhat lower than would be desired to match commercial coax.

So an additional problem beyond canceling the unwanted reactance is of matching the remaining resistive impedance to the characteristic impedance of the transmission line. In principle this can always be done with a transformer, however the turns ratio of a transformer is not adjustable. A general matching network with at least two adjustments can be made to correct both components of impedance. Matching networks using discrete inductors and capacitors will have losses associated with those components, and will have power restrictions when used for transmitting. Avoiding these difficulties, commercial antennas are generally designed with fixed matching elements or feeding strategies to get an approximate match to standard coax, such as 50 or 75 Ohms. Antennas based on the dipole (rather than vertical antennas) should include a balun in between the transmission line and antenna element, which may be integrated into any such matching network.

Another extreme case of impedance matching occurs when using a small loop antenna (usually, but not always, for receiving) at a relatively low frequency where it appears almost as a pure inductor. Resonating such an inductor with a capacitor at the frequency of operation not only cancels the reactance but greatly magnifies the very small radiation resistance of such a loop. This is implemented in most AM broadcast receivers, with a small ferrite loop antenna resonated by a capacitor which is varied along with the receiver tuning in order to maintain resonance over the AM broadcast band

Antenna Types

Antennas can be classified in various ways. The list below groups together antennas under common operating principles, following the way antennas are classified in many engineering textbooks.

Isotropic: An isotropic antenna (isotropic radiator) is a *hypothetical* antenna that radiates equal signal power in all directions. It is a mathematical model that is used as the base of comparison to calculate the gain of real antennas. No real antenna can have an isotropic radiation pattern. However *approximately* isotropic antennas, constructed with multiple elements, are used in antenna testing.

The first four groups below are usually resonant antennas; when driven at their resonant frequency their elements act as resonators. Waves of current and voltage bounce back and forth between the ends, creating standing waves along the elements.

Dipole

Yagi-Uda television antenna for analog channels 2-4, 47-68 MHz

Two-element turnstile antenna for reception of weather satellite data,
137 MHz. Has circular polarization.

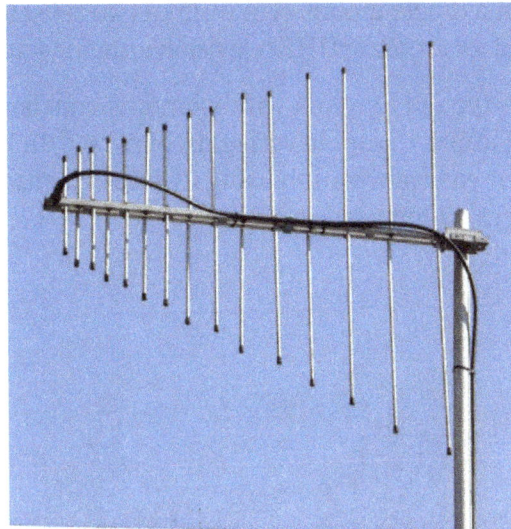

Log-periodic dipole array covering 140-470 MHz

The dipole is the prototypical antenna on which a large class of antennas are based. A basic dipole antenna consists of two conductors (usually metal rods or wires) arranged symmetrically, with one side of the balanced feedline from the transmitter or receiver attached to each. The most common type, the half-wave dipole, consists of two resonant elements just under a quarter wavelength long. This antenna radiates maximally in directions perpendicular to the antenna's axis, giving it a small directive gain of 2.15 dBi (practically the lowest directive gain of any antenna). Although half-wave dipoles are used alone as omnidirectional antennas, they are also a building block of many other more complicated directional antennas.

- *Yagi-Uda* - One of the most common directional antennas at HF, VHF, and UHF frequencies. Consists of multiple half wave dipole elements in a line, with a single driven element and multiple parasitic elements which serve to create a uni-directional or beam antenna. These typically have gains between 10 and 20 dBi depending on the number of elements used, and are very narrowband (with a usable bandwidth of only a few percent) though there are derivative designs which relax this limitation. Used for rooftop television antennas, point-to-point communication links, and long distance shortwave communication using skywave ("skip") reflection from the ionosphere.

- *Log-periodic dipole array* - Often confused with the Yagi-Uda, this consists of many dipole elements along a boom with gradually increasing lengths, all connected to the transmission line with alternating polarity. It is a directional antenna with a wide bandwidth. This makes it ideal for use as a rooftop television antenna, although its gain is much less than a Yagi of comparable size.

- *Turnstile* - Two dipole antennas mounted at right angles, fed with a phase difference of 90°. This antenna is unusual in that it radiates in *all* directions (no nodes in the radiation pattern), with horizontal polarization in directions coplaner with the elements, circular polarization normal to that plane, and elliptical polarization in other directions. Used for receiving signals from satellites, as circular polarization is transmitted by many satellites.

- *Corner reflector* - A directive antenna with moderate gain of about 8 dBi often used at UHF frequencies. Consists of a dipole mounted in front of two reflective metal screens joined at an angle, usually 90°. Used as a rooftop UHF television antenna and for point-to-point data links.

- *Patch (microstrip)* - A type of antenna with elements consisting of metal sheets mounted over a ground plane. Similar to dipole with gain of 6 - 9 dBi. Integrated into surfaces such as aircraft bodies. Their easy fabrication using PCB techniques have made them popular in modern wireless devices. Often used in arrays.

Monopole

Quarter-wave whip antenna on an FM radio for 88-108 MHz

Rubber Ducky antenna on UHF 446 MHz walkie talkie with rubber cover removed.

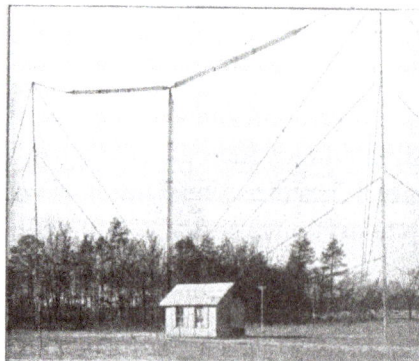

T antenna of amateur radio station, 80 ft high, used at 1.5 MHz.

Monopole antennas consist of a single conductor such as a metal rod, mounted over the ground or an artificial conducting surface (a so-called *ground plane*). One side of the feedline from the receiver or transmitter is connected to the conductor, and the other side to ground and/or the artificial ground plane. The monopole is best understood as a dipole antenna in which one conductor is omitted; the radiation is generated as if the second arm of the dipole were present due to the effective image current seen as a reflection of the monopole from the ground. Since all of the equivalent dipole's radiation is concentrated in a half-space, the antenna has twice (3 dB increase of) the gain of a similar dipole, not considering losses in the ground plane.

The most common form is the quarter-wave monopole which is one-quarter of a wavelength long and has a gain of 5.12 dBi when mounted over a ground plane. Monopoles have an omnidirectional radiation pattern, so they are used for broad coverage of an area, and have vertical polarization. The ground waves used for broadcasting at low frequencies must be vertically polarized, so large vertical monopole antennas are used for broadcasting in the MF, LF, and VLF bands. Small monopoles are used as nondirectional antennas on portable radios in the HF, VHF, and UHF bands.

- *Whip* - Type of antenna used on mobile and portable radios in the VHF and UHF bands such as boom boxes, consists of a flexible rod, often made of telescoping segments.

    o *Rubber Ducky* - Most common antenna used on portable two way radios and cordless phones due to its compactness, consists of an electrically short wire helix. The helix adds inductance to cancel the capacitive reactance of the short radiator, making it resonant. Very low gain.

    o *Ground plane* - a whip antenna with several rods extending horizontally from base of whip attached to the ground side of the feedline. Since whips are mounted above ground, the horizontal rods form an artificial ground plane under the antenna to increase its gain. Used as base station antennas for land mobile radio systems such as police, ambulance and taxi dispatchers.

- *Mast radiator* - A radio tower in which the tower structure itself serves as the antenna. Common form of transmitting antenna for AM radio stations and other MF and LF transmitters. At its base the tower is usually, but not necessarily, mounted on a ceramic insulator to isolate it from the ground.

- *T and inverted L* - Consist of a long horizontal wire suspended between two towers with insulators, with a vertical wire hanging down from it, attached to a feedline to the receiver or transmitter. Used on LF and VLF bands. The vertical wire serves as the radiator. Since at these frequencies the vertical wire is electrically short, much shorter than a quarter wavelength, the horizontal wire(s) serve as a capacitive "hat" to increase the current in the vertical radiator, increasing the gain. Very narrow bandwidth, requires loading coil to tune out the capacitive reactance and make it resonant. Requires low resistance ground (electricity)

- *Inverted F* - Combines the advantages of the inverted-L antenna and the F-type antenna of, respectively, compactness and good matching. The antenna is grounded at the base and fed at some intermediate point. The position of the feed point determines the antenna impedance. Thus, matching can be achieved without the need for an extraneous matching network.

- *Umbrella* - Very large wire transmitting antennas used on VLF bands. Consists of a central mast radiator tower attached at the top to multiple wires extending out radially from the mast to ground, like a tent or umbrella, insulated at the ends. Extremely narrow bandwidth, requires large loading coil and low resistance counterpoise ground. Used for long range military communications.

- *Collinear* - Consist of a number of dipoles in a vertical line. It is a high gain omnidirectional antenna, meaning more of the power is radiated in horizontal directions and less into the sky or ground and wasted. Gain of 8 to 10 dBi. Used as base station antennas for land mobile radio systems such as police, fire, ambulance, and taxi dispatchers, and sector antennas for cellular base stations.

- *Reflective array* - multiple dipoles in a two-dimensional array mounted in front of a flat reflecting screen. Used for radar and UHF television transmitting and receiving antennas.

- *Phased array* - A high gain antenna used at UHF and microwave frequencies which is electronically steerable. It consists of multiple dipoles in a two-dimensional array, each fed through an electronic phase shifter, with the phase shifters controlled by a computer control system. The beam can be instantly pointed in any direction over a wide angle in front of the antenna. Used for military radar and jamming systems.

- *Curtain array* - Large directional wire transmitting antenna used at HF by shortwave broadcasting stations. It consists of a vertical rectangular array of wire dipoles suspended in front of a flat reflector screen consisting of a vertical "curtain" of parallel wires, all supported between two metal towers. It radiates a horizontal beam of radio waves into the sky above the horizon, which is reflected by the ionosphere to Earth beyond the horizon

- *Batwing* or *superturnstile* - A specialized antenna used in television broadcasting consisting of perpendicular pairs of dipoles with radiators resembling bat wings. Multiple batwing antennas are stacked vertically on a mast to make VHF television broadcast antennas. Omnidirectional radiation pattern with high gain in horizontal directions. The batwing shape gives them wide bandwidth.

- *microstrip* - an array of patch antennas on a substrate fed by microstrip feedlines. Microwave antenna that can achieve large gains in compact space. Ease of fabrication by PCB techniques have made them popular in modern wireless devices. Beamwidth and polarization can be actively reconfigurable.

Loop

Loopstick antenna from an AM broadcast radio, about 4 in (10 cm) long. The antenna is inductive and, in conjunction with a variable capacitor, forms the tuned circuit at the input stage of the receiver.

Loop antennas consist of a loop or (coil) of wire. Loops with circumference of a wavelength (or integer multiple of the wavelength) are resonant and act somewhat similarly to the half-wave dipole. However a loop small in comparison to the wavelength, also called a magnetic loop, performs quite differently. This antenna interacts directly with the magnetic field of the radio wave, making it relatively insensitive to nearby electrical noise. However it has a very small radiation resistance, typically much smaller than the loss resistance, making it inefficient and thus undesirable for transmitting. They are used as receiving antennas at low frequencies, and also as direction finding antennas.

Loop antenna for transmitting at high frequencies, 2m diameter

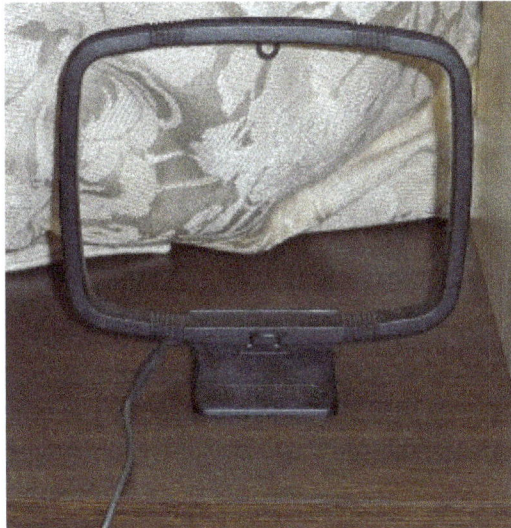

Separate loop antenna for AM radio

- Ferrite (loopstick) - These are used as the receiving antenna in most consumer AM radios operating in the medium wave broadcast band (and lower frequencies), a notable exception being car radios. Wire is coiled around a ferrite core which greatly increases the coil's inductance. Radiation pattern is maximum at directions normal to the ferrite stick.

- Quad - consists of multiple wire loops in a line with one functioning as the driven element, and the others as parasitic elements. Used as a directional antenna on the HF bands for shortwave communication.

Aperture

NASA Cassegrain parabolic spacecraft communication antenna, Australia.
Uses X band, 8 – 12 GHz. Extremely high gain ~70 dBi.

X band marine radar slot antenna on ship, 8 – 12 GHz.

Dielectric lens antenna used in millimeter wave radio telescope

Aperture antennas are the main type of directional antennas used at microwave frequencies and above. They consist of a small dipole or loop feed antenna inside a three-dimensional guiding structure large compared to a wavelength, with an aperture to emit the radio waves. Since the antenna structure itself is nonresonant they can be used over a wide frequency range by replacing or tuning the feed antenna.

- *Parabolic* - The most widely used high gain antenna at microwave frequencies and above. Consists of a dish-shaped metal parabolic reflector with a feed antenna at the focus. It can have some of the highest gains of any antenna type, up to 60 dBi, but the dish must be large compared to a wavelength. Used for radar antennas, point-to-point data links, satellite communication, and radio telescopes

- *Horn* - Simple antenna with moderate gains of 15 to 25 dBi consists of a flaring metal horn attached to a waveguide. Used for applications such as radar guns, radiometers and as feed antennas for parabolic dishes.

- *Slot* - Consist of a waveguide with one or more slots cut in it to emit the microwaves. Linear slot antennas emit narrow fan-shaped beams. Used as UHF broadcast antennas and marine radar antennas.

- *Dielectric resonator* - consists of small ball or puck-shaped piece of dielectric material excited by aperture in waveguide Used at millimeter wave frequencies

Traveling Wave

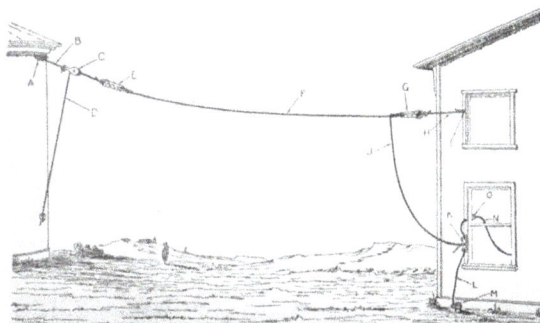

A typical random wire antenna for shortwave reception, strung between two buildings.

Quadrant antenna, similar to rhombic, at an Austrian shortwave broadcast station.
Radiates horizontal beam at 5-9 MHz, 100 kW

Unlike the above antennas, traveling wave antennas are nonresonant so they have inherently broad bandwidth. They are typically wire antennas multiple wavelengths long, through which the voltage and current waves travel in one direction, instead of bouncing back and forth to form standing waves as in resonant antennas. They have linear polarization (except for the helical antenna). Unidirectional traveling wave antennas are terminated by a resistor at one end equal to the antenna's characteristic resistance, to absorb the waves from one direction. This makes them inefficient as transmitting antennas.

- *Random wire* - This describes the typical antenna used to receive shortwave radio, consisting of a random length of wire either strung outdoors between supports or indoors in a zigzag pattern along walls, connected to the receiver at one end. Can have complex radiation patterns with several lobes at angles to the wire.

- *Beverage* - Simplest unidirectional traveling wave antenna. Consists of a straight wire one to several wavelengths long, suspended near the ground, connected to the receiver at one end and terminated by a resistor equal to its characteristic impedance, 400 to 800Ω at the other end. Its radiation pattern has a main lobe at a shallow angle in the sky off the terminated end. It is used for reception of skywaves reflected off the ionosphere in long distance "skip" shortwave communication.

- *Rhombic* - Consists of four equal wire sections shaped like a rhombus. It is fed by a balanced feedline at one of the acute corners, and the two sides are connected to a resistor equal to the characteristic resistance of the antenna at the other. It has a main lobe in a horizontal direction off the terminated end of the rhombus. Used for skywave communication on shortwave bands.

- *Helical (axial mode)* - Consists of a wire in the shape of a helix mounted above a reflecting screen. It radiates circularly polarized waves in a beam off the end, with a typical gain of 15 dBi. It is used at VHF and UHF frequencies for communication with satellites and animal tracking transmitters, which use circular polarization because it is insensitive to the relative orientation of the antennas.

- *Leaky wave* - Microwave antennas consisting of a waveguide or coaxial cable with a slot or apertures cut in it so it radiates continuously along its length.

Effect of Ground

Ground reflections is one of the common types of multipath.

The radiation pattern and even the driving point impedance of an antenna can be influenced by the dielectric constant and especially conductivity of nearby objects. For a terrestrial antenna, the ground is usually one such object of importance. The antenna's height above the ground, as well as the electrical properties (permittivity and conductivity) of the ground, can then be important. Also, in the particular case of a monopole antenna, the ground (or an artificial ground plane) serves as the return connection for the antenna current thus having an additional effect, particularly on the impedance seen by the feed line.

When an electromagnetic wave strikes a plane surface such as the ground, part of the wave is transmitted into the ground and part of it is reflected, according to the Fresnel coefficients. If the ground is a very good conductor then almost all of the wave is reflected (180° out of phase), whereas a ground modeled as a (lossy) dielectric can absorb a large amount of the wave's power. The power remaining in the reflected wave, and the phase shift upon reflection, strongly depend on the wave's angle of incidence and polarization. The dielectric constant and conductivity (or simply the complex dielectric constant) is dependent on the soil type and is a function of frequency.

For very low frequencies to high frequencies (<30 MHz), the ground behaves as a lossy dielectric, Thus the ground is characterized both by a conductivity and permittivity (dielectric constant) which can be measured for a given soil (but is influenced by fluctuating moisture levels) or can be estimated from certain maps. At lower frequencies the ground acts mainly as a good conductor, which AM middle wave broadcast (.5 - 1.6 MHz) antennas depend on.

At frequencies between 3 and 30 MHz, a large portion of the energy from a horizontally polarized antenna reflects off the ground, with almost total reflection at the grazing angles important for ground wave propagation. That reflected wave, with its phase reversed, can either cancel or reinforce the direct wave, depending on the antenna height in wavelengths and elevation angle (for a sky wave).

On the other hand, vertically polarized radiation is not well reflected by the ground except at grazing incidence or over very highly conducting surfaces such as sea water. However the grazing angle reflection important for ground wave propagation, using vertical polarization, is *in phase* with the direct wave, providing a boost of up to 6 db, as is detailed below.

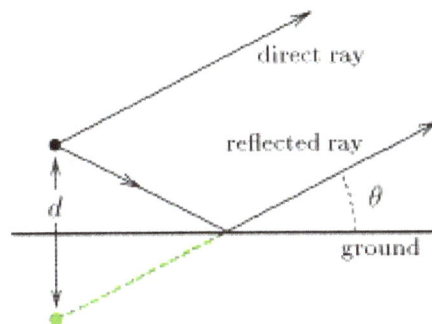

The wave reflected by earth can be considered as emitted by the image antenna.

At VHF and above (>30 MHz) the ground becomes a poorer reflector. However it remains a good reflector especially for horizontal polarization and grazing angles of incidence. That is important as these higher frequencies usually depend on horizontal line-of-sight propagation (except for satellite communications), the ground then behaving almost as a mirror.

The net quality of a ground reflection depends on the topography of the surface. When the irregularities of the surface are much smaller than the wavelength, we are in the regime of specular reflection, and the receiver sees both the real antenna and an image of the antenna under the ground due to reflection. But if the ground has irregularities not small compared to the wavelength, reflections will not be coherent but shifted by random phases. With shorter wavelengths (higher frequencies), this is generally the case.

Whenever both the receiving or transmitting antenna are placed at significant heights above the ground (relative to the wavelength), waves specularly reflected by the ground will travel a longer distance than direct waves, inducing a phase shift which can sometimes be significant. When a sky wave is launched by such an antenna, that phase shift is always significant unless the antenna is very close to the ground (compared to the wavelength).

The phase of reflection of electromagnetic waves depends on the polarization of the incident wave. Given the larger refractive index of the ground (typically $n=2$) compared to air ($n=1$), the phase of horizontally polarized radiation is reversed upon reflection (a phase shift of radians or 180°). On the other hand, the vertical component of the wave's electric field is reflected at grazing angles of incidence approximately *in phase*. These phase shifts apply as well to a ground modelled as a good electrical conductor.

The currents in an antenna appear as an image in opposite phase when reflected at grazing angles. This causes a phase reversal for waves emitted by a horizontally polarized antenna (left) but not a vertically polarized antenna (center).

This means that a receiving antenna "sees" an image of the antenna but with reversed currents. That current is in the same absolute direction as the actual antenna if the antenna is vertically oriented (and thus vertically polarized) but opposite the actual antenna if the antenna current is horizontal.

The actual antenna which is *transmitting* the original wave then also may *receive* a strong signal from its own image from the ground. This will induce an additional current in the antenna element, changing the current at the feedpoint for a given feedpoint voltage. Thus the antenna's impedance, given by the ratio of feedpoint voltage to current, is altered due to the antenna's proximity to the ground. This can be quite a significant effect when the antenna is within a wavelength or two of the ground. But as the antenna height is increased, the reduced power of the reflected wave (due to the inverse square law) allows the antenna to approach its asymptotic feedpoint impedance given by theory. At lower heights, the effect on the antenna's impedance is *very* sensitive to the exact distance from the ground, as this affects the phase of the reflected wave relative to the currents in the antenna. Changing the antenna's height by a quarter wavelength, then changes the phase of the reflection by 180°, with a completely different effect on the antenna's impedance.

The ground reflection has an important effect on the net far field radiation pattern in the vertical plane, that is, as a function of elevation angle, which is thus different between a vertically and horizontally polarized antenna. Consider an antenna at a height h above the ground, transmitting a wave considered at the elevation angle θ. For a vertically polarized transmission the magnitude of the electric field of the electromagnetic wave produced by the direct ray plus the reflected ray is:

$$|E_V| = 2|E_0|\left|\cos\left(\frac{2\pi h}{\lambda}\sin\theta\right)\right|$$

Thus the *power* received can be as high as 4 times that due to the direct wave alone (such as when θ=0), following the *square* of the cosine. The sign inversion for the reflection of horizontally polarized emission instead results in:

$$|E_H| = 2|E_0|\left|\sin\left(\frac{2\pi h}{\lambda}\sin\theta\right)\right|$$

where:

- E_0 is the electrical field that would be received by the direct wave if there were no ground.

- θ is the elevation angle of the wave being considered.

- λ is the wavelength.

- h is the height of the antenna (half the distance between the antenna and its image).

Radiation patterns of antennas and their images reflected by the ground.
At left the polarization is vertical and there is always a maximum for.
If the polarization is horizontal as at right, there is always a zero for.

For horizontal propagation between transmitting and receiving antennas situated near the ground reasonably far from each other, the distances traveled by tne direct and reflected rays are nearly the same. There is almost no relative phase shift. If the emission is polarized vertically, the two fields (direct and reflected) add and there is maximum of received signal. If the signal is polarized horizontally, the two signals subtract and the received signal is largely cancelled. The vertical plane radiation patterns are shown in the image at right. With vertical polarization there is always a maximum for θ=0, horizontal propagation (left pattern). For horizontal polarization, there is cancellation at that angle. Note that the above formulae and these plots assume the ground as a perfect conductor. These plots of the radiation pattern correspond to a distance between the antenna and its image of 2.5λ. As the antenna height is increased, the number of lobes increases as well.

The difference in the above factors for the case of θ=0 is the reason that most broadcasting (transmissions intended for the public) uses vertical polarization. For receivers near the ground, horizontally polarized transmissions suffer cancellation. For best reception the receiving antennas for these signals are likewise vertically polarized. In some applications where the receiving antenna must work in any position, as in mobile phones, the base station antennas use mixed polarization, such as linear polarization at an angle (with both vertical and horizontal components) or circular polarization.

On the other hand, classical (analog) television transmissions are usually horizontally polarized, because in urban areas buildings can reflect the electromagnetic waves and create ghost images due to multipath propagation. Using horizontal polarization, ghosting is reduced because the amount of reflection of electromagnetic waves in the p polarization (horizontal polarization off the side of a building) is generally less than s (vertical, in this case) polarization. Vertically polarized analog television has nevertheless been used in some rural areas. In digital terrestrial television such reflections are less problematic, due to robustness of binary transmissions and error correction.

Mutual Impedance and Interaction between Antennas

Current circulating in one antenna generally induces a voltage across the feedpoint of nearby antennas or antenna elements. The mathematics presented below are useful in analyzing the electrical behaviour of antenna arrays, where the properties of the individual array elements (such as half wave dipoles) are already known. If those elements were widely separated and driven in a certain amplitude and phase, then each would act independently as that element is known to. However, because of the mutual interaction between their electric and magnetic fields due to proximity, the currents in each element are *not* simply a function of the applied voltage (according to its driving point impedance), but depend on the currents in the other nearby elements. Note that this now is a near field phenomenon which could not be properly accounted for using the Friis transmission equation for instance. This near field effect creates a different set of currents at the antenna terminals resulting in distortions in the far field radiation patterns; however, the distortions may be removed using a simple set of network equations.

The elements' feedpoint currents and voltages can be related to each other using the concept of Z_{ji} between every pair of antennas just as the mutual impedance $j\omega M$ describes the voltage induced in one inductor by a current through a nearby coil coupled to it through a mutual inductance M. The mutual impedance Z_{21} between two antennas is defined as:

$$Z_{ji} = \frac{v_j}{i_i}$$

where i_i is the current flowing in antenna i and v_j is the voltage induced at the open-circuited feedpoint of antenna j due to i_1 when all other currents i_k are zero. The mutual impendances can be viewed as the elements of a symmetric square impedance matrix Z. Note that the diagonal elements, $Z_{ii} = \frac{v_i}{i_i}$, , are simply the driving point impedances of each element.

Using this definition, the voltages present at the feedpoints of a set of coupled antennas can be expressed as the multiplication of the impedance matrix times the vector of currents. Written out as discrete equations, that means:

$$
\begin{aligned}
v_1 &= i_1 Z_{11} + i_2 Z_{12} + \cdots + i_n Z_{1n} \\
v_2 &= i_1 Z_{21} + i_2 Z_{22} + \cdots + i_n Z_{2n} \\
\vdots \qquad & \quad \vdots \qquad \quad \vdots \qquad\qquad\quad \vdots \\
v_n &= i_1 Z_{n1} + i_2 Z_{n2} + \cdots + i_n Z_{nn}
\end{aligned}
$$

where:

- v_i is the voltage at the terminals of antenna i

- i_i is the current flowing between the terminals of antenna i

- Z_{ii} is the driving point impedance of antenna i

- Z_{ij} is the mutual impedance between antennas i and j..

Mutual impedance between parallel $\dfrac{\lambda}{2}$ dipoles not staggered. Curves Re and Im are the resistive and reactive parts of the impedance.

As is the case for mutual inductances,

$$
Z_{ij} = Z_{ji}.
$$

This is a consequence of Lorentz reciprocity. For an antenna element i not connected to anything (open circuited) one can write $i_i = 0$. But for an element i which is short circuited, a current is generated across that short but no voltage is allowed, so the corresponding $v_i = 0$..This is the case, for instance, with the so-called parasitic elements of a Yagi-Uda antenna where the solid rod can be viewed as a dipole antenna shorted across its feedpoint. Parasitic elements are unpowered elements that absorb and reradiate RF energy according to the induced current calculated using such a system of equations.

With a particular geometry, it is possible for the mutual impedance between nearby antennas to be zero. This is the case, for instance, between the crossed dipoles used in the turnstile antenna.

Radio Propagation

Radio propagation is the behavior of radio waves as they travel, or are propagated, from one point to another, or into various parts of the atmosphere. As a form of electromagnetic radiation, like light waves, radio waves are affected by the phenomena of reflection, refraction, diffraction, absorption, polarization, and scattering.

Radio propagation is affected by the daily changes of water vapor in the troposphere and ionization in the upper atmosphere, due to the Sun. Understanding the effects of varying conditions on radio propagation has many practical applications, from choosing frequencies for international shortwave broadcasters, to designing reliable mobile telephone systems, to radio navigation, to operation of radar systems.

Radio propagation is also affected by several other factors determined by its path from point to point. This path can be a direct line of sight path or an over-the-horizon path aided by refraction in the ionosphere, which is a region between approximately 60 and 600 km above the earth's surface. Factors influencing ionospheric radio signal propagation can include sporadic-E, spread-F, solar flares, geomagnetic storms, ionospheric layer tilts, and solar proton events.

Radio waves at different frequencies propagate in different ways. At extremely low frequencies (ELF) and very low frequencies the wavelength is much larger than the separation between the earth's surface and the D layer of the ionosphere, so electromagnetic waves may propagate in this region as a waveguide. Indeed, for frequencies below 20 kHz, the wave propagates as a single waveguide mode with a horizontal magnetic field and vertical electric field. The interaction of radio waves with the ionized regions of the atmosphere makes radio propagation more complex to predict and analyze than in free space. Ionospheric radio propagation has a strong connection to space weather. A sudden ionospheric disturbance or shortwave fadeout is observed when the x-rays associated with a solar flare ionize the ionospheric D-region. Enhanced ionization in that region increases the absorption of radio signals passing through it. During the strongest solar x-ray flares, complete absorption of virtually all ionospherically propagated radio signals in the sunlit hemisphere can occur. These solar flares can disrupt HF radio propagation and affect GPS accuracy.

Since radio propagation is not fully predictable, such services as emergency locator transmitters, in-flight communication with ocean-crossing aircraft, and some television broadcasting have been moved to communications satellites. A satellite link, though expensive, can offer highly predictable and stable line of sight coverage of a given area.

Free Space Propagation

In free space, all electromagnetic waves (radio, light, X-rays, etc.) obey the inverse-square law which states that the power density of an electromagnetic wave is proportional to the inverse of the square of the distance from a point source or:

$$\rho_P \propto \frac{1}{r^2}.$$

Doubling the distance from a transmitter means that the power density of the radiated wave at that new location is reduced to one-quarter of its previous value.

The power density per surface unit is proportional to the product of the electric and magnetic field strengths. Thus, doubling the propagation path distance from the transmitter reduces each of their received field strengths over a free-space path by one-half.

Modes

Radio frequencies and their primary mode of propagation			
Band		**Frequency**	**Wavelength**	**Propagation via**
ELF	Extremely Low Frequency	3–30 Hz	10,000-100,000 km	Fundamental mode standing wave of the Earth–ionosphere cavity with a wavelength equal to the circumference of the Earth, see Schumann resonances.
SLF	Super Low Frequency	30–300 Hz	10,000-1,000 km	
ULF	Ultra Low Frequency	0.3–3 kHz (300–3,000 Hz)	1,000–100 km	
VLF	Very Low Frequency	3–30 kHz (3,000–30,000 Hz)	100–10 km	Guided between the earth and the ionosphere.
LF	Low Frequency	30–300 kHz (30,000–300,000 Hz)	10–1 km	Guided between the earth and the D layer of the ionosphere. Surface waves.
MF	Medium Frequency	300–3000 kHz (300,000–3,000,000 Hz)	1000–100 m	Surface waves. E, F layer ionospheric refraction at night, when D layer absorption weakens.
HF	High Frequency (Short Wave)	3–30 MHz (3,000,000–30,000,000 Hz)	100–10 m	E layer ionospheric refraction. F1, F2 layer ionospheric refraction.
VHF	Very High Frequency	30–300 MHz (30,000,000–300,000,000 Hz)	10–1 m	Infrequent E ionospheric (E_s) refraction. Uncommonly F2 layer ionospheric refraction during high sunspot activity up to 50 MHz and rarely to 80 MHz. Generally direct wave. Sometimes tropospheric ducting or meteor scatter
UHF	Ultra High Frequency	300–3000 MHz (300,000,000–3,000,000,000 Hz)	100–10 cm	Direct wave. Sometimes tropospheric ducting.
SHF	Super High Frequency	3–30 GHz (3,000,000,000–30,000,000,000 Hz)	10–1 cm	Direct wave. Sometimes rain scatter.
EHF	Extremely High Frequency	30–300 GHz (30,000,000,000–300,000,000,000 Hz)	10–1 mm	Direct wave limited by absorption.
THF	Tremendously High frequency	0.3–3 THz (300,000,000,000–3,000,000,000,000 Hz)	1–0.1 mm	

Surface Modes (Groundwave)

Lower frequencies (between 30 and 3,000 kHz) have the property of following the curvature of the earth via groundwave propagation in the majority of occurrences.

In this mode the radio wave propagates by interacting with the semi-conductive surface of the earth. The wave "clings" to the surface and thus follows the curvature of the earth. Vertical polarization is used to alleviate short circuiting the electric field through the conductivity of the ground. Since the ground is not a perfect electrical conductor, ground waves are attenuated rapidly as they follow the earth's surface. Attenuation is proportional to the frequency making this mode mainly useful for LF and VLF frequencies.

Today LF and VLF are mostly used for time signals, and for military communications, especially one-way transmissions to ships and submarines, although radio amateurs have an allocation at 137 kHz in some parts of the world. Radio broadcasting using surface wave propagation uses the higher portion of the LF range in Europe, Africa and the Middle East.

Early commercial and professional radio services relied exclusively on long wave, low frequencies and ground-wave propagation. To prevent interference with these services, amateur and experimental transmitters were restricted to the higher (HF) frequencies, felt to be useless since their ground-wave range was limited. Upon discovery of the other propagation modes possible at medium wave and short wave frequencies, the advantages of HF for commercial and military purposes became apparent. Amateur experimentation was then confined only to authorized frequency segments in that range.

Direct Modes (Line-of-sight)

Line-of-sight is the direct propagation of radio waves between antennas that are visible to each other. This is probably the most common of the radio propagation modes at VHF and higher frequencies. Because radio signals can travel through many non-metallic objects, radio can be picked up through walls. This is still line-of-sight propagation. Examples would include propagation between a satellite and a ground antenna or reception of television signals from a local TV transmitter.

Ground plane reflection effects are an important factor in VHF line of sight propagation. The interference between the direct beam line-of-sight and the ground reflected beam often leads to an effective inverse-fourth-power (1/distance^4) law for ground-plane limited radiation. [Need reference to inverse-fourth-power law + ground plane. Drawings may clarify]

Ionospheric Modes (Skywave)

Skywave propagation, also referred to as skip, is any of the modes that rely on refraction of radio waves in the ionosphere, which is made up of one or more ionized layers in the upper atmosphere. F2-layer is the most important ionospheric layer for long-distance, multiple-hop HF propagation, though F1, E, and D-layers also play significant roles. The D-layer, when present during sunlight periods, causes significant amount of signal loss, as does the E-layer whose maximum usable frequency can rise to 4 MHz and above and thus block higher frequency signals from reaching the F2-layer. The layers, or more appropriately "regions", are directly affected by the sun on a daily diurnal cycle, a seasonal cycle and the 11-year sunspot cycle and determine the utility of these modes. During solar maxima, or sunspot highs and peaks, the whole HF range up to 30 MHz can be used usually around the clock and F2 propagation up to 50 MHz is observed frequently depending upon daily solar flux 10.7cm radiation values. During solar minima, or minimum sunspot counts down to zero, propagation of frequencies above 15 MHz is generally unavailable.

Although the claim is commonly made that two-way HF propagation along a given path is reciprocal, that is, if the signal from location A reaches location B at a good strength, the signal from location B will be similar at station A because the same path is traversed in both directions. However, the ionosphere is far too complex and constantly changing to support the reciprocity theorem. The path is never exactly the same in both directions. In brief, conditions at the two terminii of a path generally cause dissimilar polarization shifts, dissimilar splits into ordinary rays and extraordinary or *Pedersen rays* which are erratic and impossibly identical or similar due to variations in ionization density, shifting zenith angles, effects of the earth's magnetic DIPOLE contours, antenna radiation patterns, ground conditions and other variables.

Forecasting of skywave modes is of considerable interest to amateur radio operators and commercial marine and aircraft communications, and also to shortwave broadcasters. Real-time propagation can be assessed by listening for transmissions from specific beacon transmitters.

Meteor Scattering

Meteor scattering relies on reflecting radio waves off the intensely ionized columns of air generated by meteors. While this mode is very short duration, often only from a fraction of second to couple of seconds per event, digital Meteor burst communications allows remote stations to communicate to a station that may be hundreds of miles up to over 1,000 miles (1,600 km) away, without the expense required for a satellite link. This mode is most generally useful on VHF frequencies between 30 and 250 MHz.

Auroral Backscatter

Intense columns of Auroral ionization at 100 km altitudes within the auroral oval backscatter radio waves, perhaps most notably on HF and VHF. Backscatter is angle-sensitive—incident ray vs. magnetic field line of the column must be very close to right-angle. Random motions of electrons spiraling around the field lines create a Doppler-spread that broadens the spectra of the emission to more or less noise-like—depending on how high radio frequency is used. The radio-auroras are observed mostly at high latitudes and rarely extend down to middle latitudes. The occurrence of radio-auroras depends on solar activity (flares, coronal holes, CMEs) and annually the events are more numerous during solar cycle maxima. Radio aurora includes the so-called afternoon radio aurora which produces stronger but more distorted signals and after the Harang-minima, the late-night radio aurora (sub-storming phase) returns with variable signal strength and lesser doppler spread. The propagation range for this predominantly back-scatter mode extends up to about 2000 km in east-west plane, but strongest signals are observed most frequently from the north at nearby sites on same latitudes.

Rarely, a strong radio-aurora is followed by Auroral-E, which resembles both propagation types in some ways.

Sporadic-E Propagation

Sporadic E (Es) propagation can be observed on HF and VHF bands. It must not be confused with ordinary HF E-layer propagation. Sporadic-E at mid-latitudes occurs mostly during summer season, from May to August in the northern hemisphere and from November to February in the southern hemisphere. There is no single cause for this mysterious propagation mode. The reflection takes place in a thin sheet of ionisation around 90 km height. The ionisation patches drift westwards at speeds of few hundred km per hour. There is a weak periodicity noted during the season and typically Es is observed on 1 to 3 successive days and remains absent for a few days to reoccur again. Es do not occur during small hours; the events usually begin at dawn, and there is a peak in the afternoon and a second peak in the evening. Es propagation is usually gone by local midnight.

Observation of radio propagation beacons operating around 28.2 MHz, 50 MHz and 70 MHz, indicates that maximum observed frequency (MOF) for Es is found to be lurking around 30 MHz on most days during the summer season, but sometimes MOF may shoot up to 100 MHz or even more in ten minutes to decline slowly during the next few hours. The peak-phase includes oscillation of MOF with periodicity of approximately 5...10 minutes. The propagation range for Es single-hop is typically 1000 to 2000 km, but with multi-hop, double range is observed. The signals are very strong but also with slow deep fading.

Tropospheric Modes

Tropospheric Scattering

At VHF and higher frequencies, small variations (turbulence) in the density of the atmosphere at a height of around 6 miles (10 km) can scatter some of the normally line-of-sight beam of radio frequency energy back toward the ground, allowing over-the-horizon communication between stations as far as 500 miles (800 km) apart. The military developed the White Alice Communications System covering all of Alaska, using this tropospheric scattering principle.

Tropospheric Ducting

Sudden changes in the atmosphere's vertical moisture content and temperature profiles can on random occasions make microwave and UHF & VHF signals propagate hundreds of kilometers up to about 2,000 kilometers (1,300 mi)—and for ducting mode even farther—beyond the normal radio-horizon. The inversion layer is mostly observed over high pressure regions, but there are several tropospheric weather conditions which create these randomly occurring propagation modes. Inversion layer's altitude for non-ducting is typically found between 100 meters (300 ft) to about 1 kilometer (3,000 ft) and for ducting about 500 meters to 3 kilometers (1,600 to 10,000 ft), and the duration of the events are typically from several hours up to several days. Higher frequencies experience the most dramatic increase of signal strengths, while on low-VHF and HF the effect is negligible. Propagation path attenuation may be below free-space loss. Some of the lesser inversion types related to warm ground and cooler air moisture content occur regularly at certain times of the year and time of day. A typical example could be the late summer, early morning tropospheric enhancements that bring in signals from distances up to few hundred kilometers for a couple of hours, until undone by the Sun's warming effect.

Tropospheric Delay

This is a source of error in radio ranging techniques, such as the Global Positioning System (GPS).

Rain Scattering

Rain scattering is purely a microwave propagation mode and is best observed around 10 GHz, but extends down to a few gigahertz—the limit being the size of the scattering particle size vs. wavelength. This mode scatters signals mostly forwards and backwards when using horizontal polarization and side-scattering with vertical polarization. Forward-scattering typically yields propagation ranges of 800 km. Scattering from snowflakes and ice pellets also occurs, but scattering from ice without watery surface is less effective. The most common application for this phenomenon is microwave rain radar, but rain scatter propagation can be a nuisance causing unwanted signals to intermittently propagate where they are not anticipated or desired. Similar reflections may also occur from insects though at lower altitudes and shorter range. Rain also causes attenuation of point-to-point and satellite microwave links. Attenuation values up to 30 dB have been observed on 30 GHz during heavy tropical rain.

Airplane Scattering

Airplane scattering (or most often reflection) is observed on VHF through microwaves and, besides back-scattering, yields momentary propagation up to 500 km even in mountainous terrain. The most common back-scatter applications are air-traffic radar, bistatic forward-scatter guided-missile and airplane-detecting trip-wire radar, and the US space radar.

Lightning Scattering

Lightning scattering has sometimes been observed on VHF and UHF over distances of about 500 km. The hot lightning channel scatters radio-waves for a fraction of a second. The RF noise burst from the lightning makes the initial part of the open channel unusable and the ionization disappears quickly because of recombination at low altitude and high atmospheric pressure. Although the hot lightning channel is briefly observable with microwave radar, no practical use for this mode has been found in communications.

Other Effects

DiffractionKnife-Edge diffraction is the propagation mode where radio waves are bent around sharp edges. For example, this mode is used to send radio signals over a mountain range when a line-of-sight path is not available. However, the angle cannot be too sharp or the signal will not diffract. The diffraction mode requires increased signal strength, so higher power or better antennas will be needed than for an equivalent line-of-sight path.

Diffraction depends on the relationship between the wavelength and the size of the obstacle. In other words, the size of the obstacle in wavelengths. Lower frequencies diffract around large smooth obstacles such as hills more easily. For example, in many cases where VHF (or higher frequency) communication is not possible due to shadowing by a hill, it is still possible to communicate using the upper part of the HF band where the surface wave is of little use.

Diffraction

phenomena by small obstacles are also important at high frequencies. Signals for urban cellular telephony tend to be dominated by ground-plane effects as they travel over the rooftops of the urban environment. They then diffract over roof edges into the street, where multipath propagation, absorption and diffraction phenomena dominate.

Absorption

Low-frequency radio waves travel easily through brick and stone and VLF even penetrates sea-water. As the frequency rises, absorption effects become more important. At microwave or higher frequencies, absorption by molecular resonances in the atmosphere (mostly from water, H_2O and oxygen, O_2) is a major factor in radio propagation. For example, in the 58–60 GHz band, there is a major absorption peak which makes this band useless for long-distance use. This phenomenon was first discovered during radar research in World War II. Above about 400 GHz, the Earth's atmosphere blocks most of the spectrum while still passing some - up to UV light, which is blocked by ozone - but visible light and some of the near-infrared is transmitted. Heavy rain and falling snow also affect microwave absorption.

Measuring HF Propagation

HF propagation conditions can be simulated using radio propagation models, such as the Voice of America Coverage Analysis Program, and realtime measurements can be done using chirp transmitters. For radio amateurs the WSPR mode provides maps with real time propagation conditions between a network of transmitters and receivers. Even without special beacons the realtime propagation conditions can be measured: a worldwide network of receivers decodes morse code signals on amateur radio frequencies in realtime and provides sophisticated search functions and propagation maps for every station received.

Practical Effects

The average person can notice the effects of changes in radio propagation in several ways.

In AM broadcasting, the dramatic ionospheric changes that occur overnight in the mediumwave band drive a unique broadcast license scheme, with entirely different transmitter power output levels and directional antenna patterns to cope with skywave propagation at night. Very few stations are allowed to run without modifications during dark hours, typically only those on clear channels in North America. Many stations have no authorization to run at all outside of daylight hours. Otherwise, there would be nothing but interference on the entire broadcast band from dusk until dawn without these modifications.

For FM broadcasting (and the few remaining low-band TV stations), weather is the primary cause for changes in VHF propagation, along with some diurnal changes when the sky is mostly without cloud cover. These changes are most obvious during temperature inversions, such as in the late-night and early-morning hours when it is clear, allowing the ground and the air near it to cool more rapidly. This not only causes dew, frost, or fog, but also causes a slight "drag" on the bottom of the radio waves, bending the signals down such that they can follow the Earth's curvature over the normal radio horizon. The result is typically several stations being heard from another media market — usually a neighboring one, but sometimes ones from a few hundred kilometers away. Ice storms are also the result of inversions, but these normally cause more scattered omnidirection propagation, resulting mainly in interference, often among weather radio stations. In late spring and early summer, a combination of other atmospheric factors can occasionally cause skips that duct high-power signals to places well over 1000km away.

Non-broadcast signals are also affected. Mobile phone signals are in the UHF band, ranging from 700 to over 2600 Megahertz, a range which makes them even more prone to weather-induced propagation changes. In urban (and to some extent suburban) areas with a high population density, this is partly offset by the use of smaller cells, which use lower effective radiated power and beam tilt to reduce interference, and therefore increase frequency reuse and user capacity. However, since this would not be very cost-effective in more rural areas, these cells are larger and so more likely to cause interference over longer distances when propagation conditions allow.

While this is generally transparent to the user thanks to the way that cellular networks handle cell-to-cell handoffs, when cross-border signals are involved, unexepected charges for international roaming may occur despite not having left the country at all. This often occurs between southern San Diego and northern Tijuana at the western end of the U.S./Mexico border, and between eastern Detroit and western Windsor along the U.S./Canada border. Since signals can travel unobstructed over a body of water far larger than the Detroit River, and cool water temperatures also cause inversions in surface air, this "fringe roaming" sometimes occurs across the Great Lakes, and between islands in the Caribbean. Signals can skip from the Dominican Republic to a mountainside in Puerto Rico and vice versa, or between the U.S. and British Virgin Islands, among others. While unintended cross-border roaming is often automatically removed by mobile phone company billing systems, inter-island roaming is typically not.

References

- Lonngren, Karl Erik; Savov, Sava V.; Jost, Randy J. (2007). Fundamentals of Electomagnetics With Matlab, 2nd Ed. SciTech Publishing. p. 451. ISBN 1891121588.

- Stutzman, Warren L.; Thiele, Gary A. (2012). Antenna Theory and Design, 3rd Ed. John Wiley & Sons. pp. 560–564. ISBN 0470576642.

- Silver, H. Ward, ed. (2011). ARRL Antenna Book. Newington, Connecticut: American Radio Relay League. p. 3-2. ISBN 978-0-87259-694-8.

- Radiowave propagation, edited by M.Hall and L.Barclay, page 2, published by Peter Peregrinus Ltd., (1989), ISBN 0-86341-156-8

- Clinton B. DeSoto (1936). 200 meters & Down - The Story of Amateur Radio. W. Hartford, CT: The American Radio Relay League. pp. 132–146. ISBN 0-87259-001-1.

- Davies, Kenneth (1990). Ionospheric Radio. IEE Electromagnetic Waves Series #31. London, UK: Peter Peregrinus Ltd/The Institution of Electrical Engineers. pp. 184–186. ISBN 0-86341-186-X.

- George Jacobs and Theodore J. Cohen, Shortwave Propagation Handbook. Hicksville, New York: CQ Publishing (1982), pp. 130-135. ISBN 978-0-943016-00-9

- Chakravorty, Pragnan; Mandal, Durbadal (2016). "Radiation Pattern Correction in Mutually Coupled Antenna

Arrays Using Parametric Assimilation Technique". IEEE Transactions on Antennas and Propagation. PP (99): 1–1. doi:10.1109/TAP.2016.2578307. ISSN 0018-926X.

- Bevelaqua, Peter J. "Types of Antennas". Antenna Theory. Antenna-theory.com Peter Bevelaqua's private website. Retrieved June 28, 2015.

- Aksoy, Serkan (2008). "Antennas" (PDF). Lecture Notes-v.1.3.4. Electrical Engineering Dept., Gebze Technical University, Gebze, Turkey. Retrieved June 29, 2015.

2

Major Components of Antenna

The major components discussed in this chapter are transmitters, transposers, antenna tuners and radio receivers. A device used to rebroadcast signals in order for them to reach the receivers is known as a transposer while an antenna tuner to improves radio signals. To have a clear understanding, it's very important to comprehend the various devices of the antenna.

Transmitter

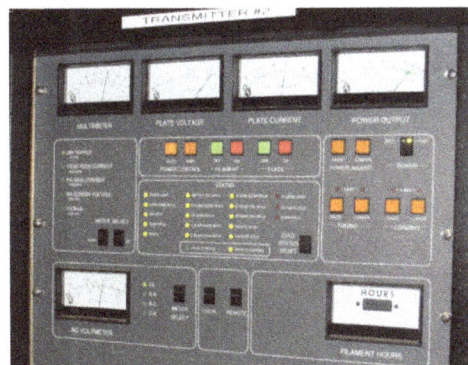

Commercial FM broadcasting transmitter at radio station WDET-FM, Wayne State University, Detroit, USA. It broadcasts at 101.9 MHz with a radiated power of 48 kW.

In electronics and telecommunications a transmitter or radio transmitter is an electronic device which, with the aid of an antenna, produces radio waves. The transmitter itself generates a radio frequency alternating current, which is applied to the antenna. When excited by this alternating current, the antenna radiates radio waves. In addition to their use in broadcasting, transmitters are necessary component parts of many electronic devices that communicate by radio, such as cell phones, wireless computer networks, Bluetooth enabled devices, garage door openers, two-way radios in aircraft, ships, spacecraft, radar sets and navigational beacons. The term *transmitter* is usually limited to equipment that generates radio waves for communication purposes; or radiolocation, such as radar and navigational transmitters. Generators of radio waves for heating or industrial purposes, such as microwave ovens or diathermy equipment, are not usually called transmitters even though they often have similar circuits.

The term is popularly used more specifically to refer to a broadcast transmitter, a transmitter used in broadcasting, as in *FM radio transmitter* or *television transmitter*. This usage typically includes both the transmitter proper, the antenna, and often the building it is housed in.

An unrelated use of the term is in industrial process control, where a "transmitter" is a telemetry device which converts measurements from a sensor into a signal, and sends it, usually via wires, to be received by some display or control device located a distance away.

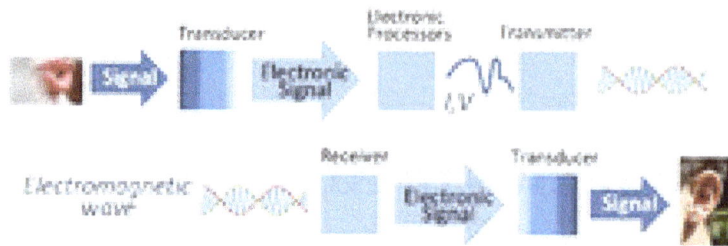

A radio transmitter is usually part of a radio communication system which uses electromagnetic waves (radio waves) to transport information (in this case sound) over a distance.

Description

A transmitter can be a separate piece of electronic equipment, or an electrical circuit within another electronic device. A transmitter and a receiver combined in one unit are called a transceiver. The term transmitter is often abbreviated "XMTR" or "TX" in technical documents. The purpose of most transmitters is radio communication of information over a distance. The information is provided to the transmitter in the form of an electronic signal, such as an audio (sound) signal from a microphone, a video (TV) signal from a video camera, or in wireless networking devices a digital signal from a computer. The transmitter combines the information signal to be carried with the radio frequency signal which generates the radio waves, which is called the carrier signal. This process is called *modulation*. The information can be added to the carrier in several different ways, in different types of transmitters. In an amplitude modulation (AM) transmitter, the information is added to the radio signal by varying its amplitude. In a frequency modulation (FM) transmitter, it is added by varying the radio signal's frequency slightly. Many other types of modulation are used.

The antenna may be enclosed inside the case or attached to the outside of the transmitter, as in portable devices such as cell phones, walkie-talkies, and garage door openers. In more powerful transmitters, the antenna may be located on top of a building or on a separate tower, and connected to the transmitter by a feed line, that is a transmission line.

Radio Transmitters

Continental 816R-5B 35 kW FM transmitter, belonging to American FM radio station KWNR broadcasting on 95.5 MHz in Las Vegas

Both the handset and the base of a cordless phone contain low power 2.4 GHz radio transmitters to communicate with each other.

remote control
outer case

remote control
inner circuit

A garage door opener control contains a low-power 2.4 GHz transmitter that sends coded commands to the garage door mechanism to open or close.

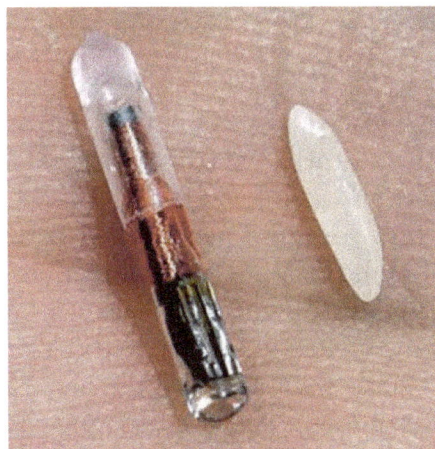

An RFID chip (next to rice grain) contains a tiny transmitter that transmits an identification number. They are incorporated into consumer products, and even implanted in pets.

History

Hertz and the first radio transmitter

The first primitive radio transmitters (called Hertzian oscillators) were built by German physicist Heinrich Hertz in 1887 during his pioneering investigations of radio waves. These generated radio waves by a high voltage spark between two conductors. Beginning in 1895 Guglielmo Marconi developed the first practical radio communication systems using spark transmitters. They couldn't transmit audio and instead transmitted information by telegraphy, the operator spelling out text messages in Morse code. These spark-gap transmitters were used during the first three decades of radio (1887-1917), called the wireless telegraphy or "spark" era. Because they generated damped waves, spark transmitters were electrically "noisy"; their energy was spread over a broad band of frequencies, creating radio noise which interfered with other transmitters. Two short-lived competing transmitter technologies came into use after the turn of the century, which were the first continuous wave transmitters: the Alexanderson alternator and Poulsen arc transmitters, which were used into the 1920s.

All these early technologies were replaced by vacuum tube transmitters in the 1920s, which used the feedback oscillator invented by Edwin Armstrong and Alexander Meissner around 1912, based on the Audion (triode) vacuum tube invented by Lee De Forest in 1906. Vacuum tube transmitters took over because they were inexpensive and produced continuous waves, which could be modulated to transmit audio (sound) using amplitude modulation (AM). This made possible commercial AM radio broadcasting, which began in about 1920. Practical frequency modulation (FM) transmission was invented by Edwin Armstrong in 1933, who showed that it was less vulnerable to noise and static than AM, and the first FM radio station was licensed in 1937. Experimental television transmission had been conducted by radio stations since the late 1920s, but practical television broadcasting didn't begin until the 1940s. The development of radar during World War 2 was a great stimulus to the evolution of high frequency transmitters in the

UHF and microwave ranges, using new devices such as the magnetron, klystron, and traveling wave tube. The invention of the transistor allowed the development in the 1960s of small portable transmitters such as wireless microphones and walkie-talkies, although the first walkie-talkie was actually produced for the military during World War 2 using vacuum tubes. In recent years, the need to conserve crowded radio spectrum bandwidth has driven the development of new types of transmitters such as spread spectrum.

Guglielmo Marconi's spark gap transmitter, with which he performed the first experiments in practical radio communication in 1895-1897

1 MW US Navy Poulsen arc transmitter which generated continuous waves using an electric arc in a magnetic field, a technology used from 1903 until the 1920s

An Alexanderson alternator, a huge rotating machine used as a radio transmitter for a short period from about 1910 until vacuum tube transmitters took over in the 1920s

How it Works

A radio transmitter is an electronic circuit which transforms electric power from a battery or electrical mains into a radio frequency alternating current, which reverses direction millions to billions of times per second. The energy in such a rapidly reversing current can radiate off a conductor (the antenna) as electromagnetic waves (radio waves). The transmitter also impresses information such as an audio or video signal onto the radio frequency current to be carried by the radio waves. When they strike the antenna of a radio receiver, the waves excite similar (but less powerful) radio frequency currents in it. The radio receiver extracts the information from the received waves. A practical radio transmitter usually consists of these parts:

- A power supply circuit to transform the input electrical power to the higher voltages needed to produce the required power output.

- An electronic oscillator circuit to generate the radio frequency signal. This usually generates a sine wave of constant amplitude often called the carrier wave, because it serves to "carry" the information through space. In most modern transmitters this is a crystal oscillator in which the frequency is precisely controlled by the vibrations of a quartz crystal.

- A modulator circuit to add the information to be transmitted to the carrier wave produced by the oscillator. This is done by varying some aspect of the carrier wave. The information is provided to the transmitter either in the form of an audio signal, which represents sound, a video signal, or for data in the form of a binary digital signal.

  o In an AM (amplitude modulation) transmitter the amplitude (strength) of the carrier wave is varied in proportion to the modulation signal.

  o In an FM (frequency modulation) transmitter the frequency of the carrier is varied by the modulation signal.

  o In an FSK (frequency-shift keying) transmitter, which transmits digital data, the frequency of the carrier is shifted between two frequencies which represent the two binary digits, 0 and 1.

  Many other types of modulation are also used. In large transmitters the oscillator and modulator together are often referred to as the *exciter*.

- An RF amplifier to increase the power of the signal, to increase the range of the radio waves.

- An impedance matching (antenna tuner) circuit to match the impedance of the transmitter to the impedance of the antenna (or the transmission line to the antenna), to transfer power efficiently to the antenna. If these impedances are not equal, it causes a condition called standing waves, in which the power is reflected back from the antenna toward the transmitter, wasting power and sometimes overheating the transmitter.

In higher frequency transmitters, in the UHF and microwave range, oscillators that operate stably at the output frequency cannot be built. In these transmitters the oscillator usually operates at a lower frequency, and is multiplied by frequency multipliers to get a signal at the desired frequency.

Legal Restrictions

In most parts of the world, use of transmitters is strictly controlled by law because of the potential for dangerous interference with other radio transmissions (for example to emergency communications). Transmitters must be licensed by governments, under a variety of license classes depending on use such as broadcast, marine radio, Airband, Amateur and are restricted to certain frequencies and power levels. A body called the International Telecommunications Union (ITU) allocates the frequency bands in the radio spectrum to various classes of users. In some classes each transmitter is given a unique call sign consisting of a string of letters and numbers which must be used as an identifier in transmissions. The operator of the transmitter usually must hold a government license, such as a general radiotelephone operator license, which is obtained by passing a test demonstrating adequate technical and legal knowledge of safe radio operation.

An exception is made allowing the unlicensed use of low-power short-range transmitters in devices such as cell phones, cordless telephones, wireless microphones, walkie-talkies, Wifi and Bluetooth devices, garage door openers, and baby monitors. In the US, these fall under Part 15 of the Federal Communications Commission (FCC) regulations. Although they can be operated without a license, these devices still generally must be type-approved before sale.

Transposer

In broadcasting, a transposer or translator is a device in or beyond the service area of a radio or television station transmitter that rebroadcasts signals to receivers which can't properly receive the signals of the transmitter because of a physical obstruction (like a hill). A translator receives the signals of the transmitter and rebroadcasts the signals to the area of poor reception. Sometimes the translator is also called a *relay transmitter, rebroadcast transmitter* or *transposer*. Since translators are used to cover a small shadowed area, their output powers are usually lower than that of the radio or television station transmitters feeding them.

Physical Obstruction

Reception of RF signals is sensitive to the size of obstruction in the path between the transmitter and the receiver. Generally speaking, if the size exceeds the wavelength the reception is interrupted. Since the wavelength is inversely proportional to frequency, it follows than that the higher frequency broadcast is more sensitive to objects between the transmitter and receiver. If the transmitter and the receiver were at the opposite sides of a hill, MW radio signals may be received, but UHF TV signals won't be received at all. That's why translators are mostly employed for VHF and UHF broadcasting (television and FM radio).

Translator Circuitry

Broadcast Station Transmitters have the Following Stages:

- Audio (AF) or video (VF) frequency buffer stages
- Modulator
- IF stages
- Mixer (IF RF)
- RF output stages (RF amplifiers and filters)

FM and TV translator stations have the following stages.

- RF input stages (RF amplifiers with AGC and band-pass filter)
- Input mixer (RF IF)
- IF stages
- Output mixer (IF RF)
- RF output stages (RF amplifiers and filters).

The output stages of both devices are similar, but the input stages are quite different. There is no baseband audio or video input to the translator. The translator receives an over-the air RF input signal by means of an antenna, just like a home receiver. Since received signal is already modulated there is no need for a modulator. Instead an input mixer or down-converter shifts the radio-frequency (RF) signal down to an intermediate-frequency (IF) signal. A second mixer (known as output mixer or up-converter) shifts the IF signal back up to the FM or TV band output signal frequency.

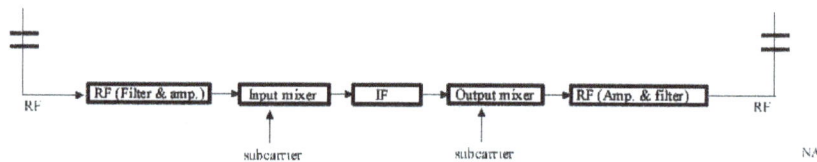

Relationship between Input and Output RF Signals.

In order to stabilize the output power, the amplification of the input RF signal is automatically controlled by PIN diodes If the frequency of the output signal were to be set equal to the frequency of the input RF signal, the output RF would feed back from the output antenna to the input antenna and the input stage would overload, completely blocking out the translator. Because of this, the translator output frequency must be different from the input signal frequency. Input and output band-pass filters further isolate the two signals.

Future of the Translator

In North America FM and TV translators were common before satellite broadcasting. With the introduction of satellite broadcasting (TVRO and RRO), some TV translator operators abandoned their stations or switched over to low power TV station (LPTV) licenses because of the higher broadcast quality provided by non-over-the-air input program streams.

With the operation of an FM or TV translator being less expensive than the same power full-service station they remained an attractive signal delivery alternative.

The transition from the analog NTSC television broadcasting standard to the digital ATSC standard resulted in a resurgence in popularity of TV translator systems in the United States.

The introduction of In-band on-channel (IBOC) hybrid analog digital FM (HDFM) technologies provided further opportunities for translator system operators.

Antenna Tuner

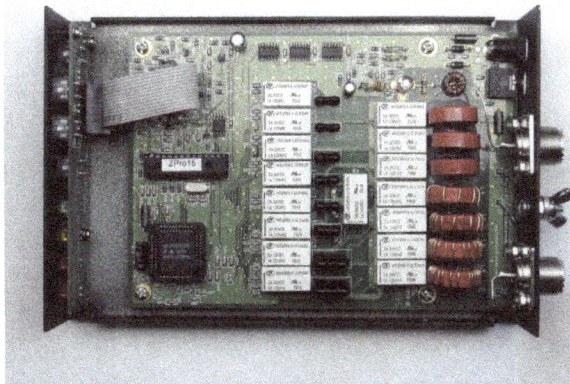

Automatic ATU for amateur transceiver

An antenna tuner, a matchbox, transmatch, antenna tuning unit (ATU), or antenna coupler is a device connected between a radio transmitter or receiver and its antenna to improve power transfer between them by matching the impedance of the radio to the antenna's feedline. Similar matching networks are used in other equipment (such as linear amplifiers) to transform impedance.

An antenna's impedance is different at different frequencies. An antenna tuner matches a radio with a fixed impedance (typically 50 Ohms for modern transceivers) to the *combination* of the feedline and the antenna; useful when the antenna's feedline impedance is unknown, complex, or otherwise different from the transceiver. Coupling through an ATU allows the use of one antenna on a broad range of frequencies. However, despite its name, an antenna '*tuner*' actually matches the transmitter only to the complex impedance reflected back to the input end of the feedline. If both tuner and transmission line were lossless, tuning at the transmitter end would produce a match at every point in the transmitter/feedline/antenna system. With practical systems, feed line losses limit the ability of the antenna 'tuner' to match the antenna or change its resonant frequency.

If the loss of power is very low in the line carrying the tranmitter's signal into the antenna, a tuner at the transmitter end can produce a worthwhile degree of matching and tuning for the antenna and feedline network as a whole. With lossy feedlines, maximum power transfer only occurs if matching is done at both ends of the line. Therefore, operating an antenna far from its design frequency and compensating with a transmatch is not as efficient as using a resonant antenna with a matched-impedance feedline. If there is still a high SWR (multiple reflections) in the feedline beyond the ATU, any loss in the feedline is multiplied several times by the transmitted waves reflecting back and forth between the tuner and the antenna, heating the wire instead of sending out a signal. Additionally, losses in the ATU itself can also waste power.

Broad Band Matching Methods

Transformers, autotransformers, and baluns are sometimes incorporated into the design of narrow band antenna tuners and antenna cabling connections. They will all usually have little effect on the resonant frequency of either the antenna or the narrow band transmitter circuits, but can widen the range of impedances that the antenna tuner can match, and/or convert between balanced and unbalanced cabling where needed.

Ferrite Transformers

Solid-state power amplifiers operating from 1–30 MHz typically use one or more wideband transformers wound on ferrite cores. MOSFETs and bipolar junction transistors are designed to operate into a low impedance, so the transformer primary typically has a single turn, while the 50 Ohm secondary will have 2 to 4 turns. This feedline system design has the advantage of reducing the retuning required when the operating frequency is changed. A similar design can match an antenna to a transmission line; For example, many TV antennas have a 300 Ohm impedance and feed the signal to the TV via a 75 Ohm coaxial line. A small ferrite core transformer makes the broad band impedance transformation. This transformer does not need, nor is it capable of adjustment. For receive-only use in a TV the small SWR variation with frequency is not a major problem.

It should be added that many ferrite based transformers perform a balanced to unbalanced transformation along with the impedance change. When the balanced to unbalanced function is present these transformers are called a balun (otherwise an unun). The most common baluns have either a 1:1 or a 1:4 *impedance* transformation.

Autotransformers

There are several designs for impedance matching using an autotransformer, which is a single-wire transformer with different connection points or taps spaced along the windings. They are distinguished mainly by their impedance transform ratio (1:1, 1:4, 1:9, etc., the square of the winding ratio), and whether the input and output sides share a common ground, or are matched from a cable that is grounded on one side (unbalanced) to an ungrounded (usually balanced) cable. When autotransformers connect balanced and unbalanced lines they are called baluns, just as two-winding transformers. When two differently-grounded cables or circuits must be connected but the grounds kept independent, a full, two-winding transformer with the desired ratio is used instead.

The circuit pictured at the right has three identical windings wrapped in the same direction around either an "air" core (for very high frequencies) or ferrite core (for middle, or low frequencies). The three equal windings shown are wired for a common ground shared by two unbalanced lines (so this design is called an unun), and can be used as 1:1, 1:4, or 1:9 impedance match, depending on the tap chosen. (The same windings could be wired differently to make a balun instead.)

For example, if the right-hand side is connected to a resistive load of 10 Ohms, the user can attach a source at any of the three ungrounded terminals on the left side of the autotransformer to get a different impedance. Notice that on the left side, the line with more windings measures greater impedance for the same 10 Ohm load on the right.

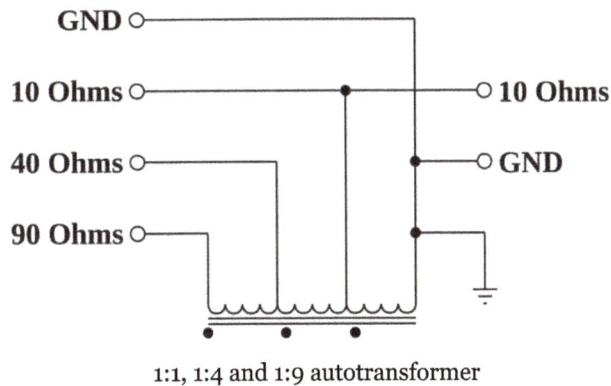

1:1, 1:4 and 1:9 autotransformer

Narrow Band Design

7Antenna matching methods that use transformers tend to cover a wide range of frequencies. A single, typical, commercially available balun can cover frequencies from 3.5–30.0 MHz, or nearly the entire shortwave radio band. Matching to an antenna using a cut segment of transmission line (described below) is perhaps the most efficient of all matching schemes in terms of electrical power, but typically can only cover a range of about 3.5–3.7 MHz – a very small range indeed, compared to a broadband balun. Antenna coupling or feedline matching circuits are also narrowband for any single setting, but can be re-tuned more conveniently. However they are perhaps the least efficient in terms of power-loss (aside from having no impedance matching at all!).

Transmission Line Antenna Tuning Methods

The insertion of a special section of transmission line, whose characteristic impedance differs from that of the main line, can be used to match the main line to the antenna. An inserted line with the proper impedance and connected at the proper location can perform complicated matching effects with very high efficiency, but spans a very limited frequency range.

The simplest example this method is the quarter-wave impedance transformer formed by a section of mismatched transmission line. If a quarter-wavelength of 75 Ohm coaxial cable is linked to a 50 Ohm load, the SWR in the 75 Ohm quarter wavelength of line can be calculated as 75Ω / 50Ω = 1.5; the quarter-wavelength of line transforms the mismatched impedance to 112.5 Ohm (75 Ohm × 1.5 = 112.5 Ohm). Thus this inserted section matches a 112 Ohm antenna to a 50 Ohm main line.

The 1/6 wavelength coaxial transformer is a useful way to match 50 to 75 ohms using the same general method. A more theoretical discussion by the inventor and wider application of the method is found here.

A second common method is the use of a stub. A shorted, or open section of line is connected in parallel with the main line. With coax this is done using a Tee connector. The length of the stub and its location can be chosen so as to produce a matched line below the stub, regardless of the complex impedance or SWR of the antenna itself. The J-pole antenna is an example of a stub matched antenna.

Basic Lumped Circuit Matching Using the L Network

The basic circuit required when lumped capacitances and inductors are used is shown below. This circuit is important in that many automatic antenna tuners use it, and also because more complex circuits can be analyzed as groups of L networks.

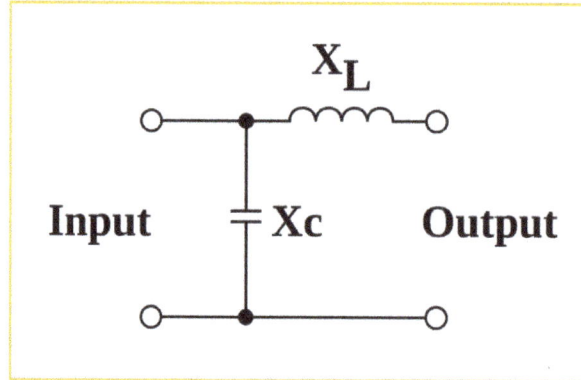

Basic network

This is called an L network not because it contains an inductor, (in fact some L networks consist of two capacitors), but because the two components are at right angles to each other, having the shape of a rotated and sometimes reversed English letter L. The T (or Tee) network and the Pi (π) networks also have a shape similar to the English and Greek letters they are named after.

This basic network is able to act as an impedance transformer. If the output has an impedance consisting of R_{load} and jX_{load}, while the input is to be attached to a source which has an impedance of R_{source} and jX_{source}

Then

$$X_L = \sqrt{\left(R_{source} + jX_{source}\right)\left(\left(R_{source} + jX_{source}\right) - \left(R_{load} + jX_{load}\right)\right)}$$

and

$$X_C = \left(R_{load} + jX_{load}\right)\sqrt{\frac{\left(R_{source} + jX_{source}\right)}{\left(R_{load} + jX_{load}\right) - \left(R_{source} + jX_{source}\right)}}$$

In this example, circuit X_L and X_c can be swapped. All the ATU circuits below create this network, which exists between systems with different impedances.

For instance, if the source has a resistive impedance of 50Ω and the load has a resistive impedance of 1000Ω:

$$X_L = \sqrt{(50)(50 - 1000)} = \sqrt{(-47500)} = j217.94 \; Ohms$$

$$X_C = 1000\sqrt{\frac{50}{(1000 - 50)}} = 1000 \times 0.2294 \; Ohms = 229.4 \; Ohms$$

If the frequency is 28 MHz,

As, $X_C = \dfrac{1}{2\pi f C}$

then, $2\pi f X_C = \dfrac{1}{C}$

So, $\dfrac{1}{2\pi f X_C} = C = 24.78 \; pF$

While as, $X_L = 2\pi f L$

then, $L = \dfrac{X_L}{2\pi f} = 1.239 \; \mu H$

Theory and Practice

A parallel network, consisting of a resistive element (1000Ω) and a reactive element (-j 229.415Ω), will have the same impedance and power factor as a series network consisting of resistive (50Ω) and reactive elements (-j 217.94Ω).

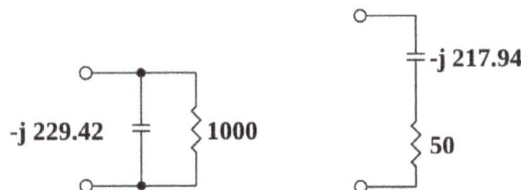

Two networks in a circuit; both have the same impedance

By adding another element in series (which has a reactive impedance of +j 217.94), the impedance is 50Ω (resistive).

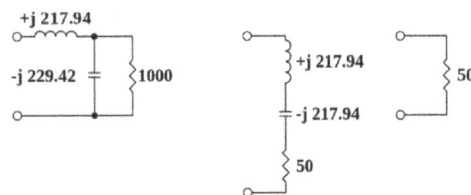

Three networks in a circuit, all with the same impedance

Types of L Networks and their Use

The L-network can have eight different configurations, six of which are shown here. The two missing configurations are the same as the bottom row, but with the parallel element (wires vertical) on the right side of the series element (wires horizontal), instead of as shown, with the parallel element on the left.

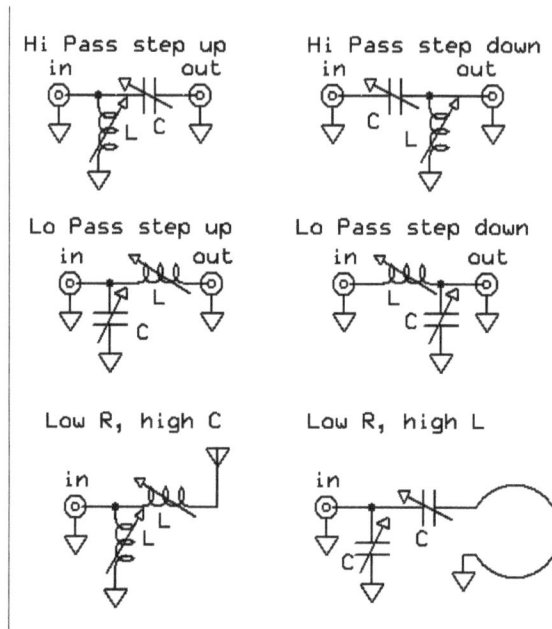

In discussion of the diagrams that follows the in connector comes from the transmitter or "source"; the out connector goes to the antenna or "load".

The general rule (with some exceptions, described below) is that the series element of an L-network goes on the side with the lowest impedance. So for example, the three circuits in the left column and the two in the bottom row have the series (horizontal) element on the out side are generally used for stepping up from a low-impedance input (transmitter) to a high-impedance output (antenna), similar to the example analyzed in the section above. The top two circuits in the right column, with the series (horizontal) element on the in side, are generally useful for stepping down from a higher input to a lower output impedance.

The general rule only applies to loads that are mainly resistive, with very little reactance. In cases where the load is highly reactive – such as an antenna fed with a signals whose frequency is far away from any resonance – the opposite configuration may be required. If far from resonance, the bottom two step down (high-in to low-out) circuits would instead be used to connect for a step up (low-in to high-out that is mostly reactance).

The low- and high-pass versions of the four circuits shown in the top two rows use only one inductor and one capacitor. Normally, the low-pass would be preferred with a transmitter, in order to attenuate harmonics, but the high-pass configuration may be chosen if the components are more conveniently obtained, or if the radio already contains an internal low-pass filter, or if attenuation of low frequencies is desirable – for example when a local AM station broadcasting on a medium frequency may be overloading a high frequency receiver.

The Low R, high C circuit is shown feeding a short vertical antenna, such as would be the case for a compact, mobile antenna or otherwise on frequencies below an antenna's lowest natural resonant frequency. Here the inherent capacitance of a short, random wire antenna is so high that the *L*-network is best realized with two inductors, instead of aggravating the problem by using a capacitor.

The Low R, high L circuit is shown feeding a small loop antenna. Below resonance tThis type of antenna has so much inductance, that more inductance from adding a coil would make the reactance even worse. Therefore the *L*-network is composed of two capacitors.

An *L*-network is the simplest circuit that will achieve the desired transformation; for any one given antenna and frequency, once a circuit is selected from the eight possible configurations (of which six are shown above) only one set of component values will match the in impedance to the out impedance. In contrast, the circuits described below all have three or more components, and hence have many more choices for inductance and capacitance that will produce an impedance match. The radio operator must exercise judgement to choose among the many adjustments that produce the same impedance match.

More Complex ATU Designs

Unbalanced Line Tuners

High Pass T Network

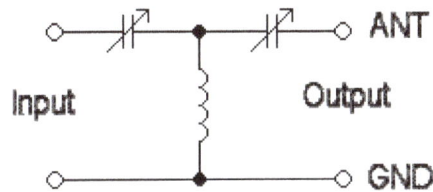

T-network transmatch

This configuration, although capable of matching a large impedance range, is a high-pass filter and will not attenuate spurious radiation above the cutoff frequency as much as the other types. Due to its low losses and simplicity, many home built and commercial manually tuned ATUs use this circuit.

Theory and Practice

If a source impedance of 200Ω and a resistive load of 1000Ω is connected (via a capacitor with an impedance of -j200Ω) to the inductor of the transmatch, vector mathematics can transform this into a parallel network consisting of a resistance of 1040 Ω and a capacitor with an admittance of 1.9231 x 10^{-4} (X$_c$ = 5200Ω).

In the following calculations, all phase angles are expressed in degrees rather than radians. A resistive load (R$_L$) of 1000Ω is in series with X$_c$ -j 200 Ω.

$$Z = \sqrt{R_L^2 + X_C^2} = 1020\Omega$$

$$\text{Phase angle}\,(\theta) = \tan^{-1}\left(\frac{X_C}{R_L}\right) = 11.31^{\circ}$$

Y = 1/Z = 9.8058 x 10^{-4}

To convert to a parallel network

$$X_C^{'} = \frac{1}{Y \sin \theta}$$

$$R_{L'} = \frac{1}{Y \cos \theta} = 1040 \ \Omega$$

If the reactive component is ignored, a 1040Ω-to-200Ω transformation is needed (according to the equations above, an inductor of +j507.32Ω). If the effect of the capacitor (from the parallel network) is taken into account, an inductor of +j462.23Ω is needed. The system can then be mathematically transformed into a series network of 199.9Ω resistive and +j409.82Ω.

A capacitor (-j409.82) is needed to complete the network. The steps are shown here. Hover over each circuit for captions.

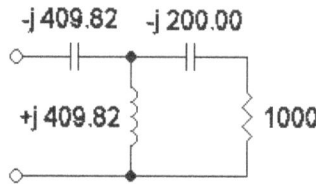

Circuit as seen by user; parts impedance shown on diagram

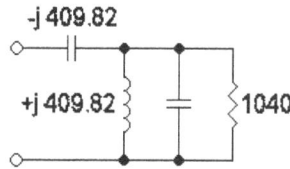

After one transformation (unlabeled part impedance is -j 5200Ω)

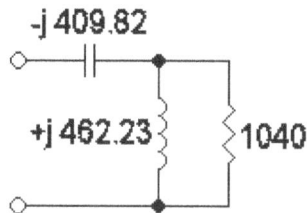

After two transformations

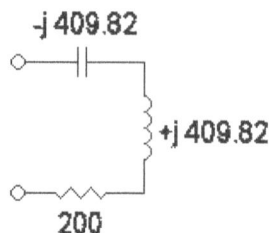

After three transformations

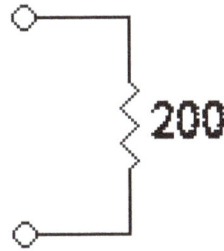

After four transformations

Π Network

A *π (pi)-network* can also be used. This ATU has very good attenuation of harmonics, but for multiband tuners the standard pi is not popular, since the variable capacitors are impractically large for the lower Amateur bands.

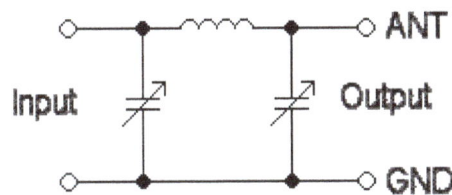

The PI Network

A modified version of the pi network is more practical as it uses a fixed input capacitor which can be several thousand picofarads while allowing the two variable capacitors to be smaller. A band switch selects the input capacitor and inductor. This circuit was used in tuners covering 1.8 to 30 MHz made by the R. L. Drake Company.

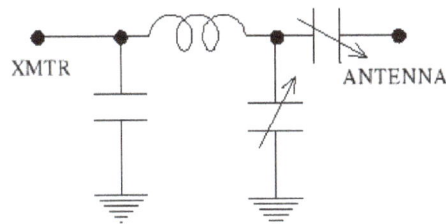

Modified pi network circuit used in Drake tuners.

Ultimate Transmatch

Originally, the *Ultimate Transmatch* was promoted as a way to make the components more manageable at the lowest frequencies of interest and also to get some harmonic attenuation. It is now considered obsolete; the design goals were better realized with the *Series-Parallel Capacitor (SPC) network*, shown below, which was designed after the name *Ultimate* had already been used. The Ultimate Transmatch network resembles an SPC tuner with the input and output reversed.

SPC Tuner

The Series Parallel Capacitor or SPC tuner can serve both as an antenna coupler and as a preselector.

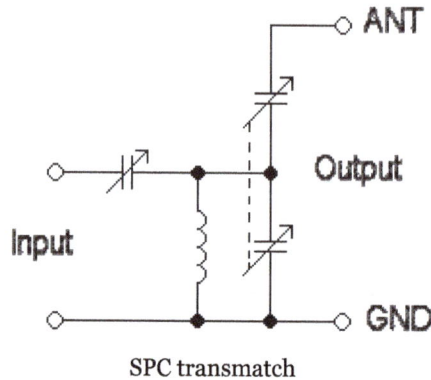

SPC transmatch

In the diagram above, the upper capacitor on the right matches impedance to the antenna, and the single capacitor on the left matches impedance to the transmitter. The coil and the lower-right capacitor form a tank circuit that drains to ground out-of-tune signals. The coil is usually also adjustable, in order to widen or narrow the band-pass and to ensure that the ganged right-hand capacitors will be able to both match to the antenna *and* tune to the transceiver's operating frequency without compromising one or the other.

Balanced Line Tuners

Balanced (open line) transmission lines require a tuner that has two "hot" output terminals, rather than one hot terminal and ground. Since all modern transmitters have unbalanced (co-axial) output, almost always 50 ohms, the most efficient system has the tuner provide a balun (balanced to unbalanced) transformation as well as providing an impedance match. The tuner usually includes a coil, and the coil can accept or produce either balanced or unbalanced input or output, depending on where the tap points are placed on the coil.

The following balanced circuit types have been used for tuners.

The match is found by tuning the capacitor and selecting taps on the main coil, which may be done with a switch accessing various taps or by physically moving clips from turn to turn. If the turns on the main coil are changed to move to a higher or lower frequency, the link turns should also change.

The Hairpin tuner has the same circuit, but uses a hairpin (transmission line) inductor. Moving the taps along the hairpin allows continuous adjustment of the impedance transformation, which is difficult with a solenoid coil. It is useful for very short wavelengths from about 10 meters to 70 cm (frequencies about 30 MHz to 430 MHz) where the solenoid inductor would have too few turns to allow fine adjustment. These tuners typically operate over at most a 2:1 frequency range.

Series Cap with Taps. Adding a series capacitor to the input side of the Fixed Link with Taps allows fine adjustment with fewer taps on the main coil. An alternate connection for the series cap circuit is useful for low impedances only, but avoids the taps (For Low Z lines on the diagram).

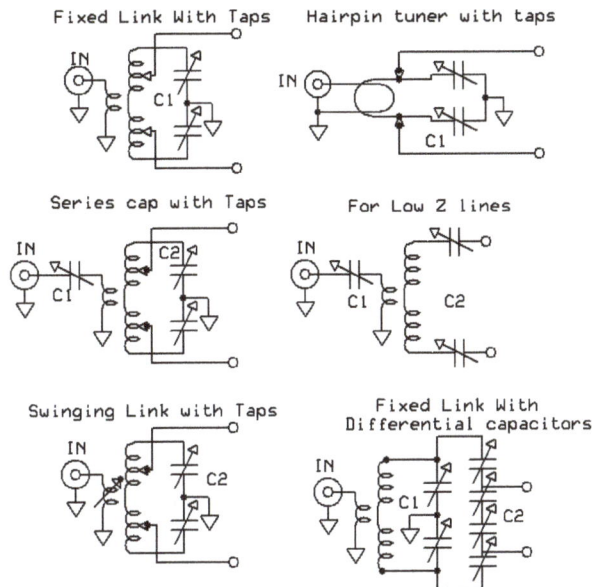

Fixed Link With Taps   Hairpin tuner with taps

Series cap with Taps   For Low Z lines

Swinging Link with Taps   Fixed Link With Differential capacitors

Referring to the diagram at the right, the Fixed Link with Taps is the most basic circuit. The factor will be nearly constant and is set by the number of relative turns on the input link.

Swinging Link with Taps. A swinging link inserted into the Fixed Link With Taps also allows fine adjustment with fewer coil taps. The *swinging link* is a form of variable transformer, that moves the input coil in and out of the space between turns in the main coil to change their mutual inductance. The variable inductance makes these tuners more flexible than the basic circuit, but at some cost in complexity.

Fixed Link with Differential Capacitors. The circuit with differential capacitors was the design used for the well-regarded Johnson Matchbox tuners. The four output capacitors (C2) are ganged, and as the top and bottom caps *increase* in value the two middle caps *decrease* in value. This provides a smooth change of loading that is equivalent to moving taps on the main coil. The Johnson Matchbox used a band switch to change the turns on the main and link inductors for each of the five frequency bands available to hams in the 1950s. The design has been criticized since the two middle-section capacitors in C2 are not strictly necessary to obtain a match; however, they conveniently reduce the disturbance that changes to C2 cause for the adjustment of C1.

Unbalanced Tuner and a Balun

Another approach to feeding balanced lines is to use an unbalanced tuner with a balun on either the input (transmitter) or output (antenna) side of the tuner. Most often using the popular high pass T circuit described above, with a 1:1 current balun on one side of the unbalanced tuner or the other.

Any balun placed on the output (antenna) side of a tuner must be built to withstand high voltage and current stresses, because of the wide range of impedances it must handle. The balun requirements are more modest when if it is put on the input (transmitter) side of the tuner, between the tuner and the receiver, since it operates at a constant impedance, but this introduces complications.

If an unbalanced tuner is fed with a balanced line from a balun instead of directly from the transmitter, then as usual its normal antenna connection – the center wire of its output coaxial cable – provides the signal to the antenna, for one side. However the ground side of that same output connection must feed an equal and opposite current to the other side of the antenna. The antenna and transmitter ground voltages lie halfway between the two "hot" feeds. In order to manage this, the common side (chassis ground) of the tuner circuit must "float", as it will be used to feed one of the hot output terminals. The high voltages present between this *floating ground* and the transmitter and antenna grounds can lead to arcing and electric shock.

To protect the operator and the equipment, and reduce power loss, an outer chassis must enclose the tuning circuit and its *floating ground*. The inner chassis can be reduced to a mounting platform within the outer chassis, elevated on insulators to separate the floating ground from contact with the other electrical grounds. In particular, the inner tuning circuit's metal mounting chassis, and the metal rods connected to adjustment knobs on the outer chassis must all be kept separate from the surface touched by the operator and from direct electrical contact with the transmitter's ground on its connection cable. This puts difficult constraints on the tuner's construction.

Z Match

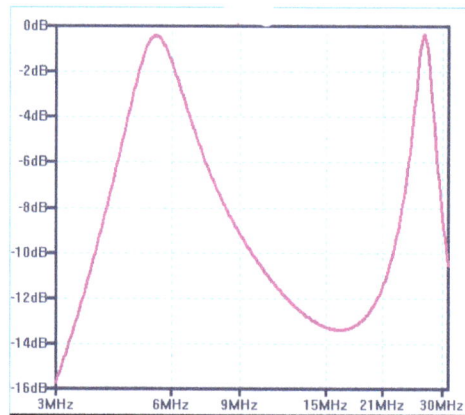

The Z match tuner response

The Z-Match is a widely used ATU in amateur radio. This tuner uses a transformer on the output side, and thus can be easily used with either balanced or unbalanced transmission lines. It is limited in power output by the core used for the output transformer. The Z match has two distinct resonant frequencies, enabling it to cover a wide frequency range without switching the inductor.

Antenna System Losses

Loss in Antenna Tuners

Every means of impedance match will introduce some power loss. This will vary from a few percent for a transformer with a ferrite core, to 50% or more for a complex ATU that is improperly tuned or working at the limits of its tuning range.

Antenna tuner front view, with partially exposed interior

With the narrow band tuners, the L network has the lowest loss, partly because it has the fewest components, but mainly because it operates at the lowest Q possible for a given impedance transformation. With The L network, the loaded Q is not adjustable, but is defined once the source and load impedances are set. Since most of the loss in practical tuners will be in the coil, choosing either the low-pass or high-pass network may reduce the loss somewhat. The L-network using only capacitors will have the lowest loss, but this network only works where the load impedance is very inductive, as occurs with a small loop antenna or a wire antenna that is between a quarter and a half wave long.

With the high-pass T network, the loss in the tuner can vary from a few percent – if tuned for lowest loss – to over 50% if the tuner is not properly adjusted. Using the maximum available capacitance will give less loss, than if one simply tunes for a match without regard for the settings. This is because using more capacitance means using less inductor turns, and the loss is mainly in the inductor.

With the SPC tuner the losses will be somewhat higher than with the T network, since the added capacitance across the inductor will shunt some reactive current to ground which must be cancelled by additional current in the inductor. The trade-off is that the effective inductance of the coil is increased, thus allowing operation at lower frequencies than would otherwise be possible.

If additional filtering is desired, the inductor can be deliberately set to larger values, thus providing a partial band pass effect. Either the high-pass T, low-pass π, or the SPC tuner can be adjusted in this manner. The additional attenuation at harmonic frequencies can be increased significantly with only a small percentage of additional loss at the tuned frequency. Note that when adjusted for minimum loss the SPC tuner will have better harmonic rejection than the high-pass T, due to the shunt capacitor. Either type is capable of good harmonic rejection if additional loss is acceptable. The low-pass π has exceptional harmonic attenuation at *any* setting, including the lowest-loss.

ATU Location

An ATU will be inserted somewhere along the line connecting the radio transmitter or receiver to the antenna. The antenna feedpoint is usually high in the air (for example, a dipole antenna) or far away (for example, an end-fed random wire antenna). A transmission line, or feedline, must carry the signal between the transmitter and the antenna. The ATU can be placed anywhere along the feedline: at the transmitter, at the antenna, or somewhere in between.

Antenna tuning is best done as close to the antenna as possible to minimize loss, increase bandwidth and reduce voltage and current on the transmission line. Also, when the information being transmitted has frequency components whose wavelength is a significant fraction of the electrical length of the feed line, distortion of the transmitted information will occur if there are standing waves on the line. Analog TV and FM stereo broadcasts are affected in this way. For those modes, matching at the antenna is required.

When possible, an automatic or remotely-controlled tuner in a weather-proof case at or near the antenna is convenient and makes for an efficient system. With such a tuner, it is possible to match a wide range of antennas (including stealth antennas).

When the ATU must be located near the radio for convent adjustment, any significant SWR will increase the loss in the feedline. For that reason, when using an ATU at the transmitter, low-loss, high-impedance feedline is a great advantage – open-wire line, for example. A short length of low-loss coaxial line is acceptable, but with longer lossy lines the additional loss due to SWR becomes very high. It is very important to remember that when matching the transmitter to the line, as is done when the ATU is near the transmitter, there is no change in the SWR in the feedline, resulting in higher loss, higher voltage or higher currents, and narrowed bandwidth, all of which are uncorrected.

Standing Wave Ratio

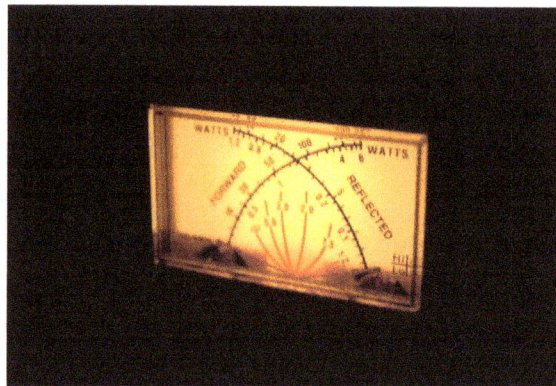

Cross-needle SWR meter on antenna tuner

It is a common misconception that a high standing wave ratio (SWR) *per se* causes loss. A well-adjusted ATU feeding an antenna through a low-loss line may have only a small percentage of additional loss compared with an intrinsically matched antenna, even with a high SWR (4:1, for example). An ATU sitting beside the transmitter just re-reflects energy reflected from the antenna (or *backlash* current) back yet again along the feedline to the antenna. High losses arise from RF resistance in the feedline and antenna, and those multiple reflections due to high SWR cause feedline losses to be compounded. Using low-loss feedline with an ATU leads to very little loss, despite multiple reflections, but a "lossy" feedline-antenna combination with an identical high SWR may loose a considerable fraction of the transmitter's power output. High impedance lines – such as most balanced lines – carry power mostly by high voltage rather than high current, so when high SWR is present there is very little loss in high-impedance line, compared low-impedance line – typical coaxial cable, for example. For that reason, radio operators are more casual about using tuners with balanced, high-impedance line.

Without an ATU, the SWR from a mismatched antenna and feedline can reflect power back into the transmitter (*backlash current*) heating and/or burning parts of the output stage. Modern solid state transmitters will automatically reduce power when high SWR is detected, so some solid-state power stages only produce weak signals if the SWR rises above 1.5 to 1. Were it not for that problem, even the losses from an SWR of 2:1 could be tolerated, since only 11 percent of transmitted power would be reflected and 89 percent sent out through to the antenna. So the main loss of output power with high SWR is due to the transmitter "backing off" its output when challenged with backlash current. Tube transmitters and amplifiers usually have an adjustable output network that can feed mismatched loads up to perhaps 3:1 SWR without trouble. In effect the built-in Pi network of the transmitter output stage acts as an ATU.

Broadcast Applications

AM Broadcast Transmitters

ATU for a 250 KW 6 tower AM Antenna

One of the oldest applications for antenna tuners is in AM and shortwave broadcasting transmitters. AM transmitters usually use a vertical antenna (tower) which can be from 0.20 to 0.68 wavelengths long. At the base of the tower an ATU is used to match the antenna to the 50 ohm transmission line from the transmitter. The most commonly used circuit is a T network, using two series inductors with a shunt capacitor between them. When multiple towers are used the ATU network may also provide for a phase adjustment so that the currents in each tower can be phased relative to the others to produce a desired pattern. These patterns are often required by law to include nulls in directions that could produce interference as well as to increase the signal in the target area. Adjustment of the ATUs in a multitower array is a complex and time consuming process requiring considerable expertise.

High-power Shortwave Transmitters

For International Shortwave (50 kW and above), frequent antenna tuning is done as part of frequency changes which may be required on a seasonal or even a daily basis. Shortwave transmitters typically include in their design the ability to match impedances up to 2:1 SWR. Modern transmitters can do this retuning within 15 seconds.

300 ohm balanced transmission lines are more or less standard for shortwave transmitters and

antennas, although other values can be found. The transmitter networks incorporate a balun or an external one is installed at the transmitter.

The most commonly used shortwave antennas for international broadcasting are the HRS antenna (curtain array), which cover a 2 to 1 frequency range and the log-periodic antenna which cover up to 8 to 1 frequency range. Within that range, the SWR will vary, but is usually kept below 1.7 to one, thus the transmitter will be able to tune itself as needed to match at any frequency.

Radio Receiver

Early broadcast radio receiver. Truetone model from about 1940

In radio communications, a radio receiver (commonly also called a radio) is an electronic device that receives radio waves and converts the information carried by them to a usable form. It is used with an antenna. The antenna intercepts radio waves (electromagnetic waves) and converts them to tiny alternating currents which are applied to the receiver, and the receiver extracts the desired information. The receiver uses electronic filters to separate the desired radio frequency signal from all the other signals picked up by the antenna, an electronic amplifier to increase the power of the signal for further processing, and finally recovers the desired information through demodulation.

The information produced by the receiver may be in the form of sound (an audio signal), images (a video signal) or digital data. A radio receiver may be a separate piece of electronic equipment, or an electronic circuit within another device. Devices that contain radio receivers include television sets, radar equipment, two-way radios, cell phones, wireless computer networks, GPS navigation devices, satellite dishes, radio telescopes, bluetooth enabled devices, garage door openers, and baby monitors.

In consumer electronics, the terms *radio* and *radio receiver* are often used specifically for receivers designed to reproduce the audio (sound) signals transmitted by radio broadcasting stations, historically the first mass-market commercial radio application.

Types of Radio Receivers

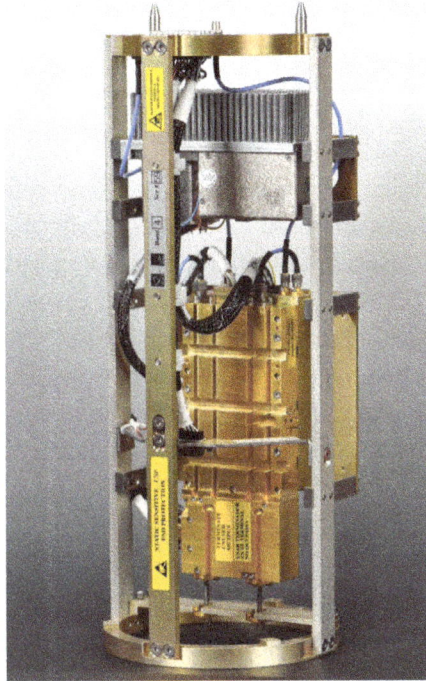

ALMA's Band 5 receivers detect electromagnetic radiation with wavelengths between about 1.4 and 1.8 mm (211 and 163 GHz). The picture shows only peripheric components of the receiver such as the Local Oscillator multiplier chain. The main receiver components of the ALMA band 5 receiver, such as horn antennae, superconductive SIS mixers, and cryogenic low-noise amplifiers, reside on a cartridge that is inserted into acryostat and cooled to 4K, 12K, and 90K, respectively

Various types of radio receivers may include:

- Consumer audio and high fidelity audio receivers and AV receivers used by home stereo listeners and audio and home theatre system enthusiasts as well as audiophiles.

- Communications receivers, used as a component of a radio communication link, characterized by high stability and reliability of performance.

- Simple crystal radio receivers, also known as a crystal set, which operate using the power received from radio waves.

- Satellite television receivers, used to receive television programming from communication satellites in geosynchronous orbit.

- Specialized-use receivers such as telemetry receivers that allow the remote measurement and reporting of information.

- Measuring receivers or measurement receivers are calibrated laboratory-grade devices that are used to measure the signal strength of broadcasting stations, the electromagnetic interference radiation emitted by electrical products, as well as to calibrate RF attenuators and signal generators.

- Scanners are specialized receivers that can automatically scan two or more discrete frequencies, stopping when they find a signal on one of them and then continuing to scan other frequencies when the initial transmission ceases. They are mainly used for monitoring VHF and UHF radio systems.

- Internet radio devices

Consumer Audio Receivers

In the context of home audio systems, the term "receiver" often refers to a combination of a tuner, a preamplifier, and a power amplifier all on the same chassis. Audiophiles will refer to such a device as an integrated receiver, while a single chassis that implements only one of the three component functions is called a discrete component. Some audio purists still prefer three discrete units - tuner, preamplifier and power amplifier - but the integrated receiver has, for some years, been the mainstream choice for music listening. The first integrated stereo receiver was made by the Harman Kardon company, and came onto the market in 1958. It had undistinguished performance, but it represented a breakthrough to the "all in one" concept of a receiver, and rapidly improving designs gradually made the receiver the mainstay of the marketplace. Many radio receivers also include a loudspeaker.

Hi-Fi / Home Theater

Today AV receivers are a common component in a high-fidelity or home-theatre system. The receiver is generally the nerve centre of a sophisticated home-theatre system providing selectable inputs for a number of different audio components like record players, CD players, tape decks and video components like VCRs, DVD players, video game consoles and television sets.

With the decline of gramophone record vinyl discs, modern receivers tend to omit inputs for phonograph turntables, which have separate requirements of their own. All other common audio/ visual components can use any of the identical line-level inputs on the receiver for playback, regardless of how they are marked (the "name" on each input is mostly for the convenience of the user). For instance, a second CD player can be plugged into an "AUX" input, and will work the same as it will in the "CD" input jacks.

Some receivers can also provide digital signal processors (DSP) to give a more realistic auditory illusion of listening in a concert hall. Digital audio S/PDIF and USB connections are also common today. The home theater receiver, in the vocabulary of consumer electronics, comprises both the 'radio receiver' and other functions, such as control, sound processing, and power amplification. The standalone radio receiver is usually known in consumer electronics as a tuner.

Some modern integrated receivers can send audio out to seven loudspeakers and an additional channel for a subwoofer and often include connections for headphones. Receivers vary greatly in price, and support stereophonic or surround sound. A high-quality receiver for dedicated audio-only listening (two channel stereo) can be relatively inexpensive. Excellent ones can be purchased for $300 or less. Because modern receivers are purely electronic devices with no moving parts unlike electromechanical devices like turntables and cassette decks, they tend to offer many years of trouble-free service. In recent years, the home theater in a box has become common, which often integrates a surround-capable receiver with a DVD player. The user simply connects it to a television, perhaps other components, and a set of loudspeakers.

Portable Radios

Early portable radio receiver

Portable radios include simple transistor radios that are typically monoaural and receive the AM, FM, or short wave broadcast bands. FM, and often AM, radios are sometimes included as a feature of portable DVD/CD, MP3 CD, and USB key players, as well as cassette player/recorders.

AM/FM stereo car radios can be a separate dashboard mounted component or a feature of in car entertainment systems.

A boombox or boom-box, sometimes known as a ghetto blaster or a jambox, in parts of Europe as a "radio-cassette", is a name given to larger portable stereo systems capable of playing radio stations and recorded music, often at a high level of volume.

Self-powered portable radios, such as clockwork radios are used in developing nations or as part of an emergency preparedness kit.

History

Radio waves were first identified in German physicist Heinrich Hertz's 1887 series of experiments to prove James Clerk Maxwell's electromagnetic theory. Hertz used micrometer spark gaps attached to dipole to produce them and loop antennas to detect them. These primitive devices are more accurately described as radio wave sensors, not "receivers", as they could only detect radio waves within about 100 feet of the transmitter, and were not used to attempt communication but were instead used as part of an experiment to verify a theoretical proof.

Spark Era

The earliest radio communication systems, used during the first three decades of radio, 1887-1917, called the *wireless telegraphy* or "spark" era, used spark gap transmitters which generated radio waves by discharging a capacitance through an electric spark. Each spark produced a transient pulse of radio waves consisting of a sinusoidal wave which decreased rapidly exponentially to zero. Spark transmitters produced strings of these damped waves, and could not generate the sinusoidal continuous waves which are modulated to carry sound in modern AM and FM transmission. Thus spark transmitters could not transmit audio (sound), and instead transmitted information by radiotelegraphy; the transmitter was switched on and off rapidly by the operator using a telegraph key, creating different length pulses of these damped radio waves ("dots" and "dashes") to spell out text messages in Morse code.

Therefore, the first radio receivers did not have to demodulate the radio signal - extract an audio signal from it as modern receivers do - just detect the presence or absence of the radio signal, and produce a sound in the earphone during the "dots" and "dashes" of the Morse code. The device which did this was called a "*detector*". Since there were no amplifying devices during this era, the sensitivity of the receiver mostly depended on the detector, and many different detector devices were tried. Radio receivers during the spark era *(see diagram)* consisted of these parts:

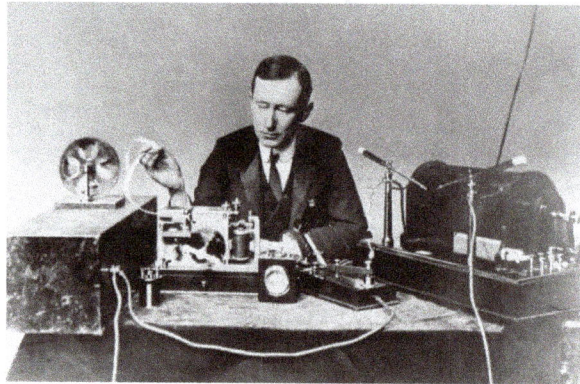

Guglielmo Marconi who built the first radio receivers, with his early spark transmitter (right) and coherer receiver (left) from the 1890s. The receiver records the Morse code on paper tape

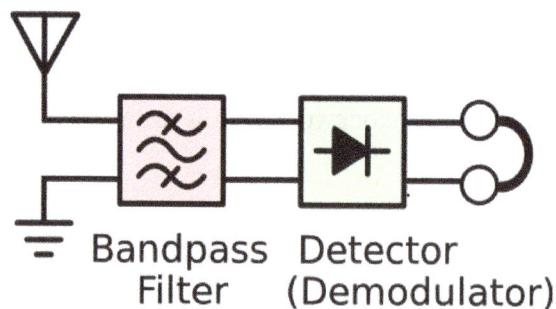

Generic block diagram of an unamplified radio receiver from the wireless telegraphy era

- An *antenna,* to intercept the radio waves and convert them to tiny radio frequency electric currents.

- A *resonant circuit* (tuned circuit), consisting of a capacitor connected to a coil of wire (an inductor), which acted as a bandpass filter to select the desired signal out of all the signals picked up by the antenna. Either the capacitor or inductor was adjustable to tune the receiver to the frequency of different transmitters. The earliest receivers, before 1897, did not have tuned circuits, they responded to all radio signals picked up by their antennas, so they had little frequency-discriminating ability and received any transmitter in their vicinity.

Most receivers used a pair of tuned circuits with their coils magnetically coupled, called a resonant transformer (oscillation transformer) or "loose coupler".

- A *detector*, which produced a pulse of DC current for each damped wave received.

- An indicating device such as an *earphone*, which converted the pulses of current into sound waves. The first receivers used an electric bell instead. Later receivers in commercial wireless systems used a Morse siphon recorder, which consisted of an ink pen mounted on a needle swung by an electromagnet (a galvanometer) which drew a line on a moving paper tape. Each string of damped waves constituting a Morse "dot" or "dash" caused the needle to swing over, which could be read off the tape. With such an automated receiver a radio operator didn't have to continuously monitor the receiver.

The signal from the spark gap transmitter consisted of damped waves repeated at an audio frequency rate, from 120 to perhaps 4000 per second, so in the earphone the signal sounded like a musical tone or buzz, and the Morse code "dots" and "dashes" sounded like beeps.

The first person to use radio waves for *communication* was Guglielmo Marconi. Marconi invented little himself, but he was first to believe that radio could be a practical communication medium, and singlehandedly developed the first wireless telegraphy systems, transmitters and receivers, beginning in 1895, mainly by improving technology invented by others. Oliver Lodge and Alexander Popov were also experimenting with similar radio wave receiving apparatus at the same time in 1895, but they are not known to have transmitted Morse code during this period, just strings of random pulses. Therefore, Marconi is usually given credit for building the first radio receivers.

Coherer Receiver

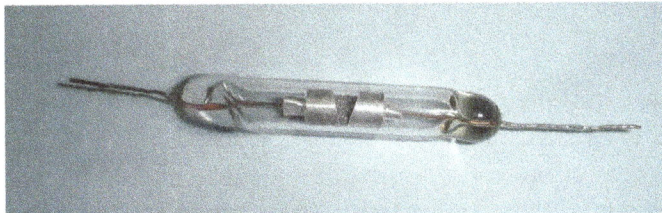

Coherer from 1904 as developed by Marconi.

One of Marconi's first coherer receivers, used in his "black box" demonstration at Toynbee Hall, London, 1896. The coherer is at right, with the "tapper" just behind it, The relay is at left, batteries are in background

A typical commercial radiotelegraphy receiver from the first decade of the 20th century.
The coherer (right) detects the pulses of radio waves, and the"dots" and "dashes" of Morse codewere recordedin
ink on paper tape by a siphon recorder (left) and transcribed later.

The first radio receivers invented by Marconi, Oliver Lodge and Alexander Popov in 1894-5 used a primitive radio wave detector called a coherer, invented in 1890 by Edouard Branly and improved by Lodge and Marconi. The coherer was a glass tube with metal electrodes at each end, with loose metal powder between the electrodes. It initially had a high resistance. When a radio frequency voltage was applied to the electrodes, its resistance dropped and it conducted electricity. In the receiver the coherer was connected directly between the antenna and ground. In addition to the antenna, the coherer was connected in a DC circuit with a battery and relay. When the incoming radio wave reduced the resistance of the coherer, the current from the battery flowed through it, turning on the relay to ring a bell or make a mark on a paper tape in a siphon recorder. In order to restore the coherer to its previous nonconducting state to receive the next pulse of radio waves, it had to be tapped mechanically to disturb the metal particles. This was done by a "decoherer", a clapper which struck the tube, operated by an electromagnet powered by the relay.

The coherer is an obscure antique device, and even today there is some disagreement about the exact physical mechanism by which the various types worked. However it can be seen that it was essentially a bistable device, a radio-wave-operated switch, and so it did not have the ability to rectify the radio wave to demodulate the later amplitude modulated (AM) radio transmissions that carried sound.

In a long series of experiments Marconi found that by using an elevated wire monopole antenna instead of Hertz's dipole antennas he could transmit longer distances, beyond the curve of the Earth, demonstrating that radio was not just a laboratory curiosity but a commercially viable communication method. This culminated in his historic transatlantic wireless transmission on December 12, 1901 from Poldhu, Cornwall to St. John's, Newfoundland, a distance of 3500 km (2200 miles), which was received by a coherer. However the usual range of coherer receivers even with the powerful transmitters of this era was limited to a few hundred miles.

The coherer remained the dominant detector used in early radio receivers for about 10 years, until replaced by the crystal detector and electrolytic detector around 1907. In spite of much development work, it was a very crude unsatisfactory device. It was not very sensitive, and also responded to impulsive radio noise (RFI), such as nearby lights being switched on or off, as well as to the intended signal. Due to the cumbersome mechanical "tapping back" mechanism it was limited to a data rate of about 12-15 words per minute of Morse code, while a spark-gap transmitter could transmit Morse at up to 100 WPM with a paper tape machine.

Other Early Detectors

The coherer's poor performance motivated a great deal of research to find better radio wave detectors, and many were invented. Some strange devices were tried; researchers experimented with using frog legs and even a human brain from a cadaver as detectors.

By the first years of the 20th century, experiments in using amplitude modulation (AM) to transmit sound by radio (radiotelephony) were being made. So a second goal of detector research was to find detectors that could demodulate an AM signal, extracting the audio (sound) signal from the radio carrier wave. It was found by trial and error that this could be done by a detector that exhibited "asymmetrical conduction"; a device that conducted current in one direction but not in the other. This rectified the alternating current radio signal, removing one side of the carrier cycles, leaving a pulsing DC current whose amplitude varied with the audio modulation signal. When applied to an earphone this would reproduce the transmitted sound.

Below are the detectors that saw wide use before vacuum tubes took over around 1920. All except the magnetic detector could rectify and therefore receive AM signals:

Magnetic detector

- Magnetic detector - Developed by Guglielmo Marconi in 1902 from a method invented by Ernest Rutherford and used by the Marconi Co. until it adopted the Audion vacuum tube around 1912, this was a mechanical device consisting of an endless band of iron wires which passed between two pulleys turned by a windup mechanism. The iron wires passed through a coil of fine wire attached to the antenna, in a magnetic field created by two magnets. The hysteresis of the iron induced a pulse of current in a sensor coil each time a radio signal passed through the exciting coil. The magnetic detector was used on shipboard receivers due to its insensitivity to vibration. One was part of the wireless station of the RMS *Titanic* which was used to summon help during its famous 15 April 1912 sinking.

- Electrolytic detector ("liquid barretter") - Invented in 1903 by Reginald Fessenden, this consisted of a thin silver-plated platinum wire enclosed in a glass rod, with the tip making contact with the surface of a cup of nitric acid. The electrolytic action caused current to be conducted in only one direction. The detector was used until about 1910. Electrolytic detectors that Fessenden had installed on US Navy ships received the first AM radio broadcast on Christmas Eve, 1906, an evening of Christmas music transmitted by Fessenden using his new alternator transmitter.

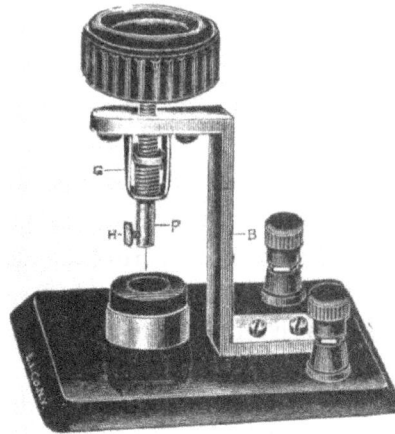

Electrolytic detector

Early Fleming valve.

Marconi valve receiver for use on ships had two Fleming valves (top) in case one burned out. It was used on the RMS Titanic.

- Thermionic diode (Fleming valve) - The first vacuum tube, invented in 1904 by John Ambrose Fleming, consisted of an evacuated glass bulb containing two electrodes: a cathode consisting of a hot wire filament similar to that in an incandescent light bulb, and a metal plate anode. Fleming, a consultant to Marconi, invented the valve as a more sensitive detector for transatlantic wireless reception. The filament was heated by a separate current through it and emitted electrons into the tube by thermionic emission, an effect which had been discovered by Thomas Edison. The radio signal was applied between the cathode and anode. When the anode was positive, a current of electrons flowed from the cathode to the anode, but when the anode was negative the electrons were repelled and no current flowed. The Fleming valve was used to a limited extent but was not popular because it was expensive, had limited filament life, and was not as sensitive as electrolytic or crystal detectors.

A galena cat's whisker detector from a 1920s crystal radio

- Crystal detector (cat's whisker detector) - invented around 1904-1906 by Henry H. C. Dunwoody and Greenleaf Whittier Pickard, based on Karl Ferdinand Braun's 1874 discovery of "asymmetrical conduction" in crystals, these were the most successful and widely used detectors before the vacuum tube era and gave their name to the *crystal radio* receiver *(below)*. One of the first semiconductor electronic devices, a crystal detector consisted of a pea-sized pebble of a crystalline semiconductor mineral such as galena (lead sulfide) whose surface was touched by a fine springy metal wire mounted on an adjustable arm. This functioned as a primitive diode which conducted electric current in only one direction. In addition to their use in crystal radios, carborundum crystal detectors were also used in some early vacuum tube radios because they were more sensitive than the vacuum tube grid-leak detector.

During the vacuum tube era, the term "detector" changed from meaning a radio wave detector to mean a demodulator, a device that could extract the audio modulation signal from a radio signal. That is its meaning today.

Tuning

The word "tuning" means adjusting the frequency of the receiver (its *passband*) to the frequency of the desired radio transmitter, to receive the desired radio transmission. Although in many modern radio receivers (such as television sets) this is done automatically by pressing a "channel up" or "channel down" button, some older radios still use the traditional method of turning a "tuning" knob until the station is heard in the speaker. However the very first receivers in the spark era could not be tuned at all.

The first receivers had no tuned circuit, the detector was connected directly between the antenna and ground. Due to the lack of any frequency selective components besides the antenna, the bandwidth of the receiver was equal to the broad bandwidth of the antenna. This was acceptable and even necessary because the first Hertzian spark transmitters also lacked a tuned circuit. Due to the impulsive nature of the spark, they radiated a very "noisy" signal, the energy of the radio waves was spread over a very wide band of frequencies, a wide bandwidth. To receive enough energy from this wideband signal the receiver had to have a wide bandwidth also.

It was found that when more than one spark transmitter was radiating in a given area, their frequencies overlapped, so their signals interfered with each other, resulting in garbled reception. It became clear that, if multiple transmitters were to operate simultaneously, some method of *selective signaling* was needed, to allow the receiver to select which transmitter's signal to receive. As radio transmission and detection systems were being developed it was also theorized that the multiple wavelengths produced by a poorly tuned transmitter were overlapping each other causing the signal to "dampen", or die down, greatly reducing the power and range of transmission. In 1892, William Crookes gave a lecture on radio in which he suggested using resonance to reduce the bandwidth of transmitters and receivers. Different transmitters could then be "tuned" to transmit on different frequencies so they didn't interfere. The receiver would also have a resonant circuit (tuned circuit), and could receive a particular transmission by "tuning" its resonant circuit to the same frequency as the transmitter, analogously to tuning a musical instrument to resonance with another. This is the system used in all modern radio.

Tuning was used in Hertz's original experiments and practical application of tuning showed up in the early to mid 1890's in wireless systems not specifically designed for radio communication. Nikola Tesla's March 1893 lecture demonstrating the wireless transmission of power for lighting (mainly by what he thought was ground conduction) included elements of tuning. The wireless lighting system consisted of a spark-excited grounded resonant transformer with a wire antenna which transmitted power across the room to another resonant transformer tuned to the frequency of the transmitter, which lighted a Geissler tube. Use of tuning in free space "Hertzian waves" (radio) was explained and demonstrated in Oliver Lodge's 1894 lectures on Hertz's work. At the time Lodge was demonstrating the physics and optical qualities of radio waves instead of attempting to build a communication system but he would go on to develop methods (patented in 1897) of tuning radio (what he called "syntony"), including using variable inductance to tune antennas.

By 1897 the advantages of tuned systems had become clear, and Marconi and the other wireless researchers had incorporated tuned circuits, consisting of capacitors and inductors connected together, into their transmitters and receivers. The tuned circuit acted like an electrical analog of a tuning fork. It had a high impedance at its resonant frequency, but a low impedance at all other frequencies. Connected between the antenna and the detector it served as a bandpass filter, passing the signal of the desired station to the detector, but routing all other signals to ground. The frequency of the station received f was determined by the capacitance C and inductance L in the tuned circuit:

$$f = \frac{1}{2\pi\sqrt{LC}}$$

Inductive Coupling

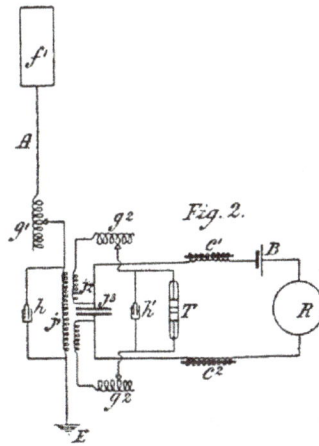

Marconi's inductively coupled coherer receiver from his controversial April 1900 "four circuit" patent no. 7,777.

Braun receiving transformer from 1904

Crystal receiver from 1914 with "loose coupler" tuning transformer. The secondary coil (1) can be slid in or out of the primary (in box) to adjust the coupling. Other components: (2) primary tuning capacitor, (3) secondary tuning capacitor, (4) loading coil, (5) crystal detector, (8) headphones

In order to reject radio noise and interference from other transmitters near in frequency to the desired station, the bandpass filter (tuned circuit) in the receiver has to have a narrow bandwidth, allowing only a narrow band of frequencies through. The form of bandpass filter that was used in the first receivers, which has continued to be used in receivers until recently, was the double-tuned inductively-coupled circuit, or resonant transformer (oscillation transformer or RF transformer). The antenna and ground were connected to a coil of wire, which was magnetically coupled to a second coil with a capacitor across it, which was connected to the detector. The RF alternating current from the antenna through the primary coil created a magnetic field which induced a current in the secondary coil which fed the detector. Both primary and secondary were tuned circuits; the primary coil resonated with the capacitance of the antenna, while the secondary coil resonated with the capacitor across it. Both were adjusted to the same resonant frequency.

This circuit had two advantages. One was that by using the correct turns ratio, the impedance of the antenna could be matched to the impedance of the receiver, to transfer maximum RF power to the receiver. Impedance matching was important to achieve maximum receiving range in the unamplified receivers of this era. The coils usually had taps which could be selected by a multiposition switch. The second advantage was that due to "loose coupling" it had a much narrower bandwidth than a simple tuned circuit, and the bandwidth could be adjusted. Unlike in an ordinary transformer, the two coils were "loosely coupled"; separated physically so not all the magnetic field from the primary passed through the secondary, reducing the mutual inductance. This gave the coupled tuned circuits much "sharper" tuning, a narrower bandwidth than a single tuned circuit. In the "Navy type" loose coupler, widely used with crystal receivers, the smaller secondary coil was mounted on a rack which could be slid in or out of the primary coil, to vary the mutual inductance between the coils. When the operator encountered an interfering signal at a nearby frequency, the secondary could be slid further out of the primary, reducing the coupling, which narrowed the bandwidth, rejecting the interfering signal. A disadvantage was that all three adjustments in the loose coupler - primary tuning, secondary tuning, and coupling - were interactive; changing one changed the others. So tuning in a new station was a process of successive adjustments.

Selectivity became more important as spark transmitters were replaced by continuous wave transmitters which transmitted on a narrow band of frequencies, and broadcasting led to a proliferation of closely spaced radio stations crowding the radio spectrum. Resonant transformers continued to be used as the bandpass filter in vacuum tube radios, and new forms such as the *variometer* were invented. Another advantage of the double-tuned transformer for AM reception was that when properly adjusted it had a "flat top" frequency response curve as opposed to the "peaked" response of a single tuned circuit. This allowed it to pass the sidebands of AM modulation on either side of the carrier with little distortion, unlike a single tuned circuit which attenuated the higher audio frequencies. Until recently the bandpass filters in the superheterodyne circuit used in all modern receivers were made with resonant transformers, called IF transformers.

Patent Disputes

Marconi's initial radio system had relatively poor tuning limiting its range and adding to interference. To overcome this drawback he developed a four circuit system with tuned coils in "*symphony*" at both the transmitters and receivers. His 1900 British #7,777 (four sevens) patent for tuning filed in April 1900 and granted a year later opened the door to patents disputes since it infringed on the Syntonic patents of Oliver Lodge, first filed in May of 1897, as well as patents filed by Ferdinand Braun. Marconi was able to obtain patents in the UK and France but the US version of his tuned four circuit patent, filed in November 1900, was initially rejected based on it being anticipated by Lodge's tuning system, and refiled versions were rejected because of the prior patents by Braun, and Lodge. A further clarification and re-submission was rejected because it infringed on parts of two prior patents Tesla had obtained for his wireless power transmission system. Marconi's lawyers manged to get a resubmitted patent reconsidered by another examiner who initially rejected it due to a pre-existing John Stone Stone tuning patent, but it was finally approved it in June of 1904 based on it having a unique system of variable inductance tuning that was different from Stone who tuned by varying the length of the antenna. When Lodge's Syntonic patent was extended in 1911 for another 7 years the Marconi Company agreed to settle that patent dispute, purchasing Lodge's radio company with its patent in 1912, giving them the priority patent they needed. Other patent disputes would crop up over the years including a 1943 US Supreme Court ruling on the Marconi Companies ability to sue the US government over patent infringement during World War One. The Court rejected the Marconi Companies suit saying they could not sue for patent infringement when their own patents did not seem to have priority over the patents of Lodge, Stone, and Tesla.

Crystal Radio Receiver

Prior to 1920 the crystal receiver was the main type used in wireless telegraphy stations, and sophisticated models were made, like this Marconi Type 106 from 1915.

Family listening to the first broadcasts around 1920 with a crystal receiver.
The mother and father have to share an earphone

After vacuum tube receivers appeared around 1920, the crystal set became a simple cheap alternative radio used by youth and the poor.

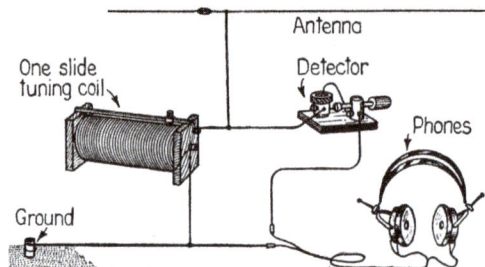

Simple crystal radio. The capacitance of the wire antenna connected to the coil serves as the capacitor in the tuned circuit.

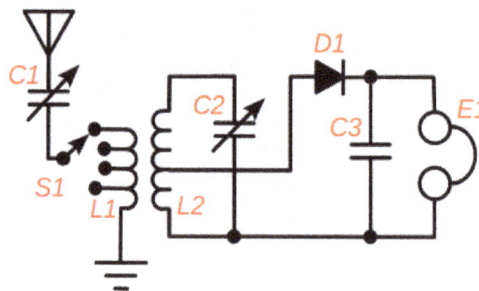

Typical "loose coupler" crystal radio circuit

Although it was invented in 1904 in the wireless telegraphy era, the crystal radio receiver could also rectify AM transmissions and served as a bridge to the broadcast era. In addition to being the main type used in commercial stations during the wireless telegraphy era, it was the first receiver to be used widely by the public. During the first two decades of the 20th century, radio listening became a popular hobby, which accelerated as stations began to transmit in AM voice (radiotelephony) instead of telegraphy, and the crystal was the simplest cheapest detector. The millions of people who purchased or homemade these inexpensive reliable receivers created the mass listening audience for the first radio broadcasts, which began around 1920. Superseded by vacuum tube receivers in the 1920s, it continued to be used by youth and the poor until World War 2. Today crystal radios, the simplest type of radio receivers, are constructed by students as educational science projects.

The crystal radio used a crystal detector called a cat's whisker detector, invented by Harrison H. C. Dunwoody and Greenleaf Whittier Pickard in 1904, to rectify the radio signal to extract the audio from the radio frequency carrier. It consisted of a mineral crystal, usually galena (PbS, lead sulfide) which was lightly touched by a fine springy wire (the "cat whisker") on an adjustable arm. The resulting crude semiconductor junction functioned as a Schottky barrier diode, it only conducted current in one direction. Only particular sites on the crystal surface functioned as rectifying junctions, and the junction could be disrupted by the slightest vibration. So a usable site was found by trial and error before each use; the operator would drag the cat's whisker across the crystal until the radio began functioning.

Like other receivers of this era the crystal radio was unamplified and ran off the power of the radio waves received from the radio station, so it had to be listened to with earphones; it could not drive a loudspeaker. It required a long wire antenna, and its sensitivity depended on how large the antenna was. During the wireless era it was used in commercial and military longwave stations with huge antennas to receive long distance radiotelegraphy traffic, even including transatlantic traffic. However, when used to receive broadcast stations a typical home crystal set had a more limited range of 25 – 100 miles. In sophisticated crystal radios the "loose coupler" inductively-coupled tuned circuit was used to increase the Q. However it still had poor selectivity compared to modern receivers.

Heterodyne Receiver and BFO

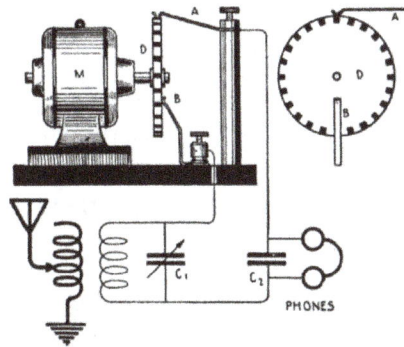

Radio receiver with Poulsen "tikker" consisting of a commutator disk turned by a motor to interrupt the carrier.

Beginning around 1905 continuous wave (CW) transmitters began to replace spark transmitters for radiotelegraphy because they had much greater range. The first continuous wave transmitters were the Poulsen arc invented in 1904 and the Alexanderson alternator developed 1906-1910, which were replaced by vacuum tube transmitters beginning around 1920.

The continuous wave radiotelegraphy signals produced by these transmitters required a different method of reception. The radiotelegraphy signals produced by spark gap transmitters consisted of strings of damped waves repeating at an audio rate, so the "dots" and "dashes" of Morse code were audible as a tone or buzz in the receivers' earphones. However the new continuous wave radiotelegraph signals simply consisted of pulses of unmodulated carrier (sine waves). These were inaudible in the receiver headphones. To receive this new modulation type, the receiver had to produce some kind of tone during the pulses of carrier.

The first crude device that did this was the "ticker" or "tikker", invented in 1908 by Valdemar Poulsen. This was a vibrating interrupter with a capacitor at the tuner output which served as a rudimentary modulator, interrupting the carrier at an audio rate, thus producing a buzz in the earphone when the carrier was present. A similar device was the "tone wheel" invented by Rudolph Goldschmidt, a wheel spun by a motor with contacts spaced around its circumference, which made contact with a stationary brush.

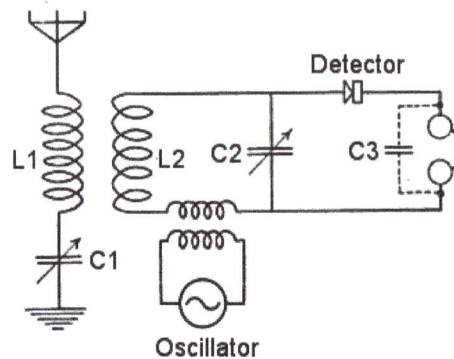

Fessenden's heterodyne radio receiver circuit

In 1901 Reginald Fessenden had invented a better means of accomplishing this. In his *heterodyne receiver* an unmodulated sine wave radio signal at a frequency f_0 offset from the incoming radio wave carrier f_c was applied to a rectifying detector such as a crystal detector or electrolytic detector, along with the radio signal from the antenna. In the detector the two signals mixed, creating two new *heterodyne* (beat) frequencies at the sum $f_c + f_0$ and the difference $f_c - f_0$ between these frequencies. By choosing f_0 correctly the lower heterodyne $f_c - f_0$ was in the audio frequency range, so it was audible as a tone in the earphone whenever the carrier was present. Thus the "dots" and "dashes" of Morse code were audible as musical "beeps". A major attraction of this method during this pre-amplification period was that the heterodyne receiver actually amplified the signal somewhat, the detector had "mixer gain".

The receiver was ahead of its time, because when it was invented there was no oscillator capable of producing the radio frequency sine wave f_0 with the required stability. Fessenden first used his large radio frequency alternator, but this wasn't practical for ordinary receivers. The heterodyne receiver remained a laboratory curiosity until a cheap compact source of continuous waves appeared, the vacuum tube electronic oscillator invented by Edwin Armstrong and Alexander Meissner in 1913. After this it became the standard method of receiving CW radiotelegraphy. The heterodyne oscillator is the ancestor of the *beat frequency oscillator* (BFO) which is used to receive radiotelegraphy in communications receivers today. The heterodyne oscillator had to be retuned each time the receiver was tuned to a new station, but in modern superheterodyne receivers the BFO signal beats with the fixed intermediate frequency, so the beat frequency oscillator can be a fixed frequency.

Armstrong later used Fessenden's heterodyne principle in his superheterodyne receiver *(below)*.

Vacuum Tube Era

During the "Golden Age of Radio" (1920 to 1950), families gathered to listen to the home radio in the evening, such as this Zenith console model 12-S-568 from 1938, a 12 tube superheterodyne with pushbutton tuning and 12 inch cone speaker.

The Audion (triode) vacuum tube invented by Lee De Forest in 1906 was the first practical amplifying device and revolutionized radio. Vacuum tube transmitters replaced spark transmitters and made possible four new types of modulation: continuous wave (CW) radiotelegraphy, amplitude modulation (AM) around 1915 which could carry audio (sound), frequency modulation (FM) around 1938 which had much improved audio quality, and single sideband (SSB).

SUPER-HETERODYNE

THE TROPADYNE

THE NEUTRODYNE

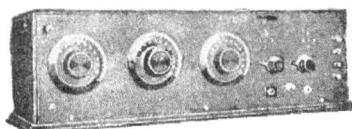

5-TUBE COCKADAY

THE REFLEX

Unlike today, when almost all radios use a variation of the superheterodyne design, during the 1920s vacuum tube radios used a variety of competing circuits.

The amplifying vacuum tube used energy from a battery or electrical outlet to increase the power of the radio signal, so vacuum tube receivers could be more sensitive and have a greater reception range than the previous unamplified receivers. The increased audio output power also allowed them to drive loudspeakers instead of earphones, permitting more than one person to listen. The first loudspeakers were produced around 1915. These changes caused radio listening to evolve explosively from a solitary hobby to a popular social and family pastime. The development of amplitude modulation (AM) and vacuum tube transmitters during World War 1, and the availability of cheap receiving tubes after the war, set the stage for the start of AM broadcasting, which sprang up spontaneously around 1920.

The advent of radio broadcasting increased the market for radio receivers greatly, and transformed them into a consumer product. At the beginning of the 1920s the radio receiver was a forbidding high-tech device, with many cryptic knobs and controls requiring technical skill to operate, housed in an unattractive black metal box, with a tinny-sounding horn loudspeaker. By the 1930s, the broadcast receiver had become a piece of furniture, housed in an attractive wooden case, with standardized controls anyone could use, which occupied a respected place in the home living room. In the early radios the multiple tuned circuits required multiple knobs to be adjusted to tune in a new station. One of the most important ease-of-use innovations was "single knob tuning", achieved by linking the tuning capacitors together mechanically. The dynamic cone loudspeaker invented in 1924 greatly improved audio frequency response over the previous horn speakers, allowing music to be reproduced with good fidelity. Convenience features like large lighted dials, tone controls, pushbutton tuning, tuning indicators and automatic gain control (AGC) were added. The receiver market was divided into the above *broadcast receivers* and *communications receivers*, which were used for two-way radio communications such as shortwave radio.

A vacuum tube receiver required several power supplies at different voltages, which in early radios were supplied by separate batteries. These were the standard batteries used in early radios:

- "A" battery - This supplied current to heat the filaments of the tubes, which consumed the bulk of the power in early radios. The first tubes used 6V at several amperes, so lead-acid automobile batteries were often used, as they could be recharged. Later tubes used 3V or 1.5V from dry cell batteries.

- "B" battery - this supplied the plate (anode) voltage for the tubes, including the audio output power to the earphone or loudspeaker. These were rectangular multicell carbon-zinc batteries. They were made in multiples of 22.5 volts: 22.5, 45, 67.5, and 90 volts, and often had taps to give different voltages.

- "C" battery - a few radios required a third voltage of about 4V to bias the grid of the tubes negative.

By 1930 adequate rectifier tubes were developed, and the expensive batteries were replaced by a transformer power supply that worked off the house current.

Vacuum tubes were bulky, expensive, had a limited lifetime, consumed a large amount of power and produced a lot of waste heat, so the number of tubes a receiver could economically have was a limiting factor. Therefore, a goal of tube receiver design was to get the most performance out of a limited number of tubes. The major radio receiver designs, listed below, were invented during the vacuum tube era.

A defect in many early vacuum tube receivers was that the amplifying stages could oscillate, act as an oscillator, producing unwanted radio frequency alternating currents. These parasitic oscillations mixed with the carrier of the radio signal in the detector tube, producing audible beat notes (heterodynes); annoying whistles, moans, and howls in the speaker. This was due to feedback in the amplifiers; one major feedback path was the capacitance between the plate and grid in early triodes. This was solved by the Neutrodyne circuit, and later the development of the tetrode and pentode around 1930.

Edwin Armstrong is one of the most important figures in radio receiver history, and during this period invented technology which continues to dominate radio communication. He was the first to give a correct explanation of how De Forest's triode tube worked. He invented the feedback oscillator, regenerative receiver, the superregenerative receiver, the superheterodyne receiver, and modern frequency modulation (FM).

The First Vacuum Tube Receivers

De Forest's first commercial Audion receiver, the RJ6 which came out in 1914. The Audion tube was always mounted upside down, with its delicate filament loop hanging down, so it did not sag and touch the other electrodes in the tube.

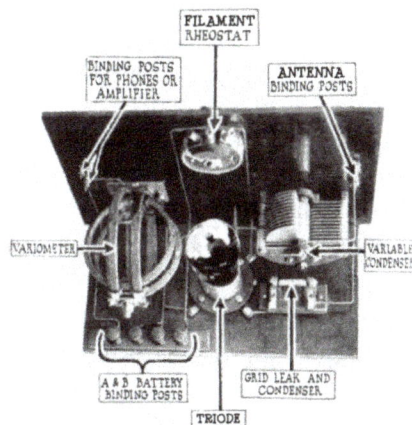

The first amplifying vacuum tube, the Audion, a crude triode, was invented in 1906 by Lee De Forest as a more sensitive detector for radio receivers, by adding a third electrode to the thermionic diode detector, the Fleming valve. It was not widely used until its amplifying ability was recognized around 1912. The first tube receivers, invented by De Forest and built by hobbyists until the mid 1920s, used a single Audion which functioned as a grid-leak detector which both rectified and amplified the radio signal. The grid-leak detector circuit was also used in regenerative, TRF, and early superheterodyne receivers (below) until the 1930s.

To give enough output power to drive a loudspeaker, 2 or 3 additional Audion stages were needed for audio amplification. Many early hobbyists could only afford a single tube receiver, and listened to the radio with earphones, so early tube amplifiers and speakers were sold as add-ons.

In addition to very low gain of about 5 and a short lifetime of about 100 hours, the primitive Audion had erratic characteristics because it was incompletely evacuated, some residual air was left in the tube by De Forest, who believed that ionization was key to its operation. This made it a more sensitive detector but also caused its electrical characteristics to vary during use. As the tube heated up, gas released from the metal elements would change the pressure in the tube, changing the plate current and other characteristics, so it required periodic bias adjustments to keep it at the correct operating point. Each Audion stage usually had a rheostat to adjust the filament current, and often a potentiometer or multiposition switch to control the plate voltage. The filament rheostat was also used as a volume control. The many controls made multitube Audion receivers nightmarishly complicated to operate.

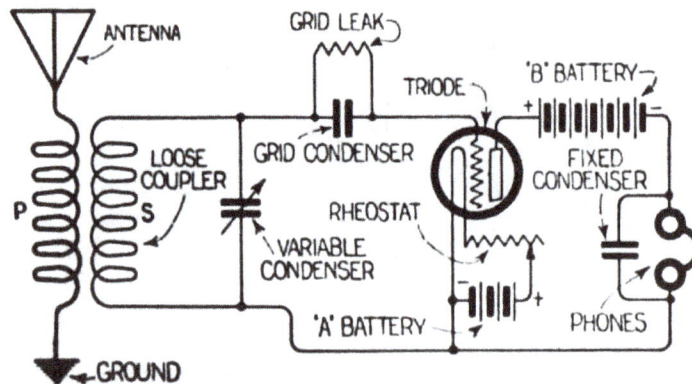

Example of single tube triode grid-leak receiver from 1920, the first type of amplifying radio receiver.
In the grid leak circuit, electrons attracted to the grid during the positive half cycles of the radio signalcharge the grid capacitor with a negative voltage of a few volts, biasing the grid near its cutoff voltage, so the tube conducts only during the positive half-cycles,rectifying the radio carrier.

By 1914, Harold Arnold at Western Electric and Irving Langmuir at GE had realized that the residual gas in the tube that caused these problems was not necessary; the Audion could operate on electron conduction alone. They were able to evacuate tubes to a lower pressure of 10^{-9} atm, producing the first "hard vacuum" triodes. These more stable tubes did not require bias adjustments and allowed radios to have fewer controls and be more user-friendly. During World War 1 civilian radio use was prohibited, but by 1920 these tubes came on the market and large scale production of vacuum tube radios began. The "soft" incompletely-evacuated tubes were used as detectors through the 1920s then became obsolete.

Regenerative (Autodyne) Receiver

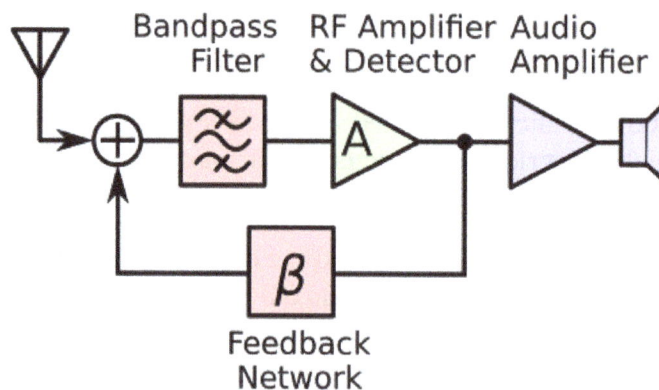

Block diagram of regenerative receiver

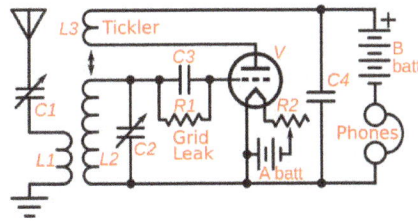

Circuit of single tube Armstrong regenerative receiver

Homemade Armstrong regenerative receiver, 1922. The "tickler" coil *(L3)* is visible on the front panel, coupled to the input tuning coils.

Commercial regenerative receiver from the early 1920s, the Paragon RA-10 *(center)* with separate 10R single tube RF amplifier *(left)* and three tube DA-2 detector and 2-stage audio amplifier unit *(right)*. The 4 cylindrical dry cell "A" batteries *(right rear)* powered the tube filaments, while the 2 rectangular "B" batteries provided plate voltage.

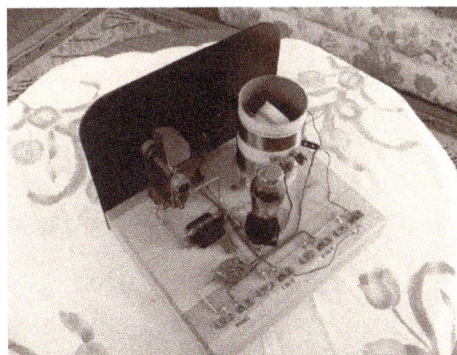

Homemade one-tube Armstrong regenerative receiver from the 1940s. The ticklercoil is a variometer winding mounted on a shaft inside the tuning coil (upper right) which can be rotated by a knob on the front panel.

The regenerative receiver, invented by Edwin Armstrong in 1913 when he was a 23-year-old college student, was used very widely until the late 1920s particularly by hobbyists who could only afford a single-tube radio. Today transistor versions of the circuit are still used in a few inexpensive applications like walkie-talkies. In the regenerative receiver the gain (amplification) of a vacuum tube or transistor is increased by using *regeneration* (positive feedback); some of the energy from the tube's output circuit is fed back into the input circuit with a feedback loop. The early vacuum tubes had very low gain (around 5). Regeneration could not only increase the gain of the tube enormously, by a factor of 15,000 or more, it also increased the Q factor of the tuned circuit, decreasing (sharpening) the bandwidth of the receiver by the same factor, improving selectivity greatly. The receiver had a control to adjust the feedback. The tube also acted as a grid-leak detector to rectify the AM signal.

Another advantage of the circuit was that the tube could be made to oscillate, and thus a single tube could serve as both a beat frequency oscillator and a detector, functioning as a heterodyne receiver to make CW radiotelegraphy transmissions audible. This mode was called an autodyne receiver. To receive radiotelegraphy, the feedback was increased until the tube oscillated, then the oscillation frequency was tuned to one side of the transmitted signal. The incoming radio carrier signal and local oscillation signal mixed in the tube and produced an audible heterodyne (beat) tone at the difference between the frequencies.

A widely used design was the Armstrong circuit, in which a "tickler" coil in the plate circuit was coupled to the tuning coil in the grid circuit, to provide the feedback. The feedback was controlled by a variable resistor, or alternately by moving the two windings physically closer together to increase loop gain, or apart to reduce it. This was done by an adjustable air core transformer called a variometer (variocoupler). Regenerative detectors were sometimes also used in TRF and superheterodyne receivers.

One problem with the regenerative circuit was that when used with large amounts of regeneration the selectivity (Q) of the tuned circuit could be *too* sharp, attenuating the AM sidebands, thus distorting the audio modulation. This was usually the limiting factor on the amount of feedback that could be employed.

A more serious drawback was that it could act as an inadvertent radio transmitter, producing interference (RFI) in nearby receivers. In AM reception, to get the most sensitivity the tube was operated very close to instability and could easily break into oscillation (and in CW reception *did* oscillate), and the resulting radio signal was radiated by its wire antenna. In nearby receivers, the regenerative's signal would beat with the signal of the station being received in the detector, creating annoying heterodynes, (beats), howls and whistles. Early regeneratives which oscillated easily were called "bloopers", and were made illegal in Europe. One preventative measure was to use a stage of RF amplification before the regenerative detector, to isolate it from the antenna. But by the mid 1920s "regens" were no longer sold by the major radio manufacturers.

Superregenerative Receiver

This was a receiver invented by Edwin Armstrong in 1922 which used regeneration in a more sophisticated way, to give greater gain. It was used in a few shortwave receivers in the 1930s, and is used today in a few cheap high frequency applications such as walkie-talkies and garage door openers.

In the regenerative receiver the loop gain of the feedback loop was less than one, so the tube (or other amplifying device) did not oscillate but was close to oscillation, giving large gain. In the superregenerative receiver, the loop gain was made equal to one, so the amplifying device actually began to oscillate, but the oscillations were interrupted periodically. This allowed a single tube to produce gains of over 10^6.

Tuned Radio Frequency (TRF) Receiver

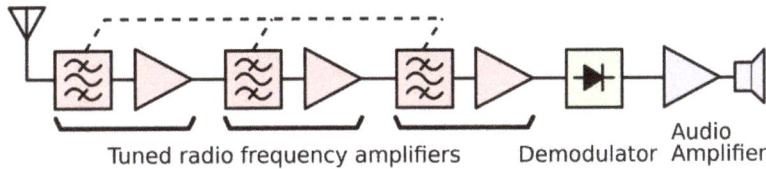

Block diagram of a tuned radio frequency receiver. The dotted line indicates that the bandpass filters must be tuned together.

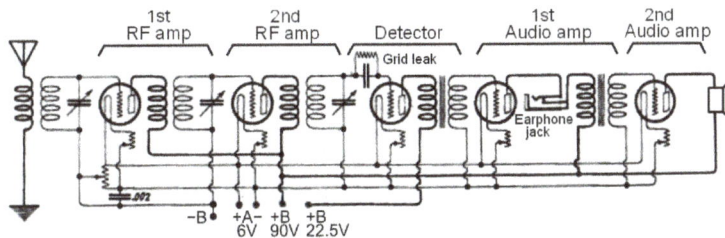

Typical 5 tube TRF circuit from 1924 has 2 stages of RF amplification, a grid-leak detector stage, and 2 stages of transformer-coupled audio amplification

Early 6 tube TRF receiver from around 1920. The 3 large knobs adjust the 3 tuned circuits to tune in stations

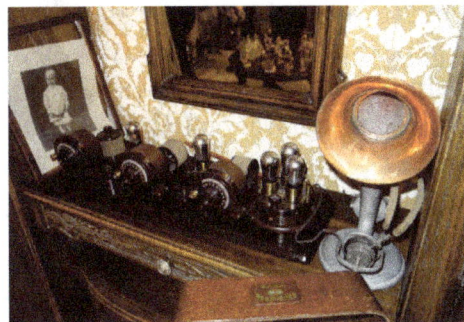

Atwater-Kent TRF receiver from the 1920s with 2 RF stages (left), detector and two audio amplifier tubes (right). The loudspeaker consists of an earphone coupled to an acoustic horn which amplifies the sound.

Tuning a Neutrodyne TRF receiver with 3 tuned circuits *(large knobs)*

The tuned radio frequency (TRF) receiver, invented in 1916 by Ernst Alexanderson, improved both sensitivity and selectivity by using several stages of amplification before the detector, each with a tuned circuit, all tuned to the frequency of the station. It was very popular in quality radios during the 1920s until the superheterodyne replaced it in the 1930s. The TRF receiver consisted of these parts:

- One or more tuned radio frequency amplifier stages, each consisting of an amplifying tube or transistor and a tuned circuit. In vacuum tube radios these consisted of a tube amplifier followed by an air core interstage coupling transformer with a capacitor across one winding.

- A detector stage, in tube radios usually a triode grid-leak detector.

- One or more audio amplifier stages

A major problem of early TRF receivers was that they were complicated to tune, because each resonant circuit had to be adjusted to the frequency of the station before the radio would work. In later TRF receivers the tuning capacitors were linked together mechanically ("ganged") on a common shaft so they could be adjusted with one knob, but in early receivers the frequencies of the tuned circuits could not be made to "track" well enough to allow this, and each tuned circuit had its own tuning knob. Therefore, the knobs had to be turned simultaneously. For this reason most TRF sets had no more than three tuned RF stages.

A second problem was that the multiple radio frequency stages, all tuned to the same frequency, were prone to oscillate, and the parasitic oscillations mixed with the radio station's carrier in the detector, producing audible heterodynes (beat notes), whistles and moans, in the speaker. This was solved by the invention of the Neutrodyne circuit *(below)* and the development of the tetrode later around 1930, and better shielding between stages.

Today the TRF design is used in a few integrated (IC) receiver chips. From the standpoint of modern receivers the disadvantage of the TRF is that the gain and bandwidth of the tuned RF stages are not constant but vary as the receiver is tuned to different frequencies. Since the bandwidth of a filter with a given Q is proportional to the frequency, as the receiver is tuned to higher frequencies its bandwidth increases.

Neutrodyne Receiver

The Neutrodyne receiver, invented in 1922 by Louis Hazeltine, was a TRF receiver with a "neutralizing" circuit added to each radio amplification stage to cancel the feedback to prevent the oscillations which caused the annoying whistles in the TRF. In the neutralizing circuit a capacitor fed a feedback current from the plate circuit to the grid circuit which was 180° out of phase with the feedback which caused the oscillation, canceling it. The Neutrodyne was popular until the advent of cheap tetrode tubes around 1930.

Reflex Receiver

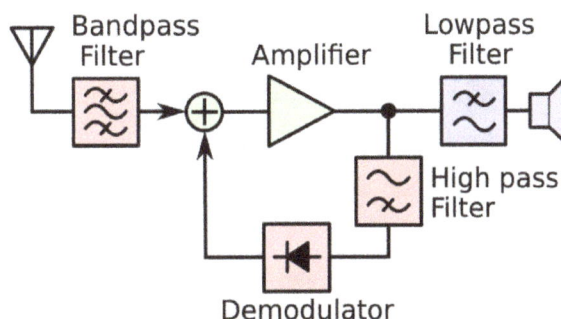

Block diagram of simple single tube reflex receiver

The reflex receiver, invented in 1914 by Wilhelm Schloemilch and Otto von Bronk, and rediscovered and extended to multiple tubes in 1917 by Marius Latour and William H. Priess, was a design used in some inexpensive radios of the 1920s which enjoyed a resurgence in small portable tube radios of the 1930s and again in a few of the first transistor radios in the 1950s. It is another example of an ingenious circuit invented to get the most out of a limited number of active devices. In the reflex receiver the RF signal from the tuned circuit is passed through one or more amplifying tubes or transistors, demodulated in a detector, then the resulting audio signal is passed *again* though the same amplifier stages for audio amplification. The separate radio and audio signals present simultaneously in the amplifier do not interfere with each other since they are at different frequencies, allowing the amplifying tubes to do "double duty". In addition to single tube reflex receivers, some TRF and superheterodyne receivers had several stages "reflexed". Reflex radios were prone to a defect called "play-through" which meant that the volume of audio did not go to zero when the volume control was turned down.

Superheterodyne Receiver

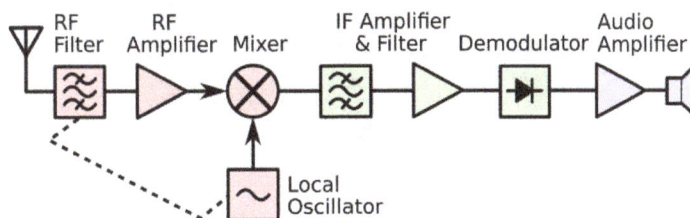

Block diagram of a superheterodyne receiver. The dotted line indicates that the RF filter and local oscillator must be tuned in tandem.

The first superheterodyne receiver built at Armstrong's Signal Corps laboratory in Paris during World War I.
It is constructed in two sections, the mixer and local oscillator (left) and three IF amplification
stages anda detector stage (right). The intermediate frequency was 75 kHz.

During the 1940s the vacuum tube superheterodyne receiver was refined into acheap-to-manufacture
form called the "All American Five" because it only required 5 tubes,which was used in
almost all broadcast radios until the end of the tube era in the 1970s.

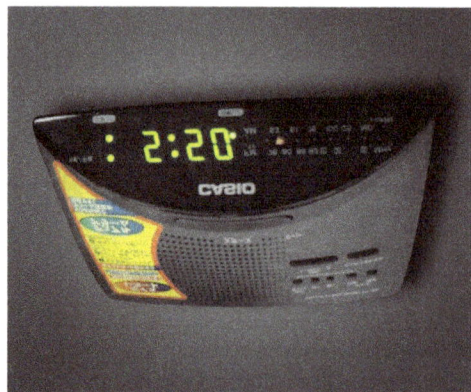

Modern transistor superheterodyne clock radio

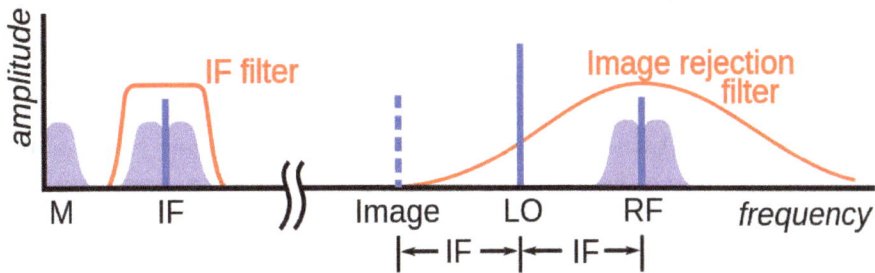

How a superheterodyne works. The incoming radio signal (RF) from the antenna consists of a carrier frequency (dark blue) with sidebands (light blue) on either side containing the modulation. Mixing this with the local oscillator signal (LO), creates a heterodyne intermediate frequency (IF) at the difference between these frequencies. This is bandpass filtered in the IF amplifier, and the demodulator extracts the modulation (M). Because the Image frequency when mixed with the LO also creates a signal at the IF, the RF image rejection filter on the input of the receiver filters out any radio signal at the image frequency.

The superheterodyne, invented in 1918 during World War 1 by Edwin Armstrong when he was in the Signal Corps, is the design used in almost all modern receivers, except a few specialized applications. It is a more complicated design than the other receivers above, and when it was invented required 6 - 9 vacuum tubes, putting it beyond the budget of most consumers, so it was initially used mainly in commercial and military communication stations. However, by the 1930s the "superhet" had replaced all the other receiver types above.

In the superheterodyne, the "heterodyne" technique invented by Reginald Fessenden is used to shift the frequency of the radio signal down to a lower "intermediate frequency" (IF), before it is processed. An unmodulated sinusoidal signal f_{LO} generated by a *local oscillator* (LO) is mixed with the incoming radio signal f_{RF} from the antenna in a nonlinear vacuum tube or transistor called the "*mixer*". The result at the output of the mixer is a heterodyne or beat frequency at the difference between these two frequencies: $f_{IF} = |f_{RF} - f_{LO}|$ This lower frequency is called the *intermediate frequency* (IF). The IF also contains the modulation (sidebands) that was present in the original RF signal. This signal is amplified and bandpass filtered as in the TRF receiver, then is demodulated in a detector, producing an audio signal.

The superheterodyne design is superior to other receivers for these reasons:

- At the high frequencies used for radio transmission, signal processing circuitry often performs poorly. Amplifying devices have little gain, and are prone to instability and parasitic oscillation, as was seen in the TRF receiver. By shifting the signal down to a lower intermediate frequency, the amplification, filtering and detection can be done at a more convenient frequency where the electronics works better. Armstrong invented the superheterodyne to receive high frequencies around 5 MHz, which were above the 1 MHz frequency range of most vacuum tubes of the time. Today active devices still perform poorly at the microwave frequencies which modern telecommunication links use.

- The different frequencies of different stations are all converted to the same frequency, the IF, for filtering, so the bandwidth and gain of the receiver is constant over its frequency range. To tune the receiver to a different frequency, only the frequency of the local oscillator fLO needs to be changed. The rest of the receiver after the mixer operates at a fixed frequency, the IF. All the other receivers above require that the bandpass filter (tuned circuit) be adjustable to different frequencies. Since the bandwidth of the filter is proportional to the frequency, the bandwidth of the receiver increases as it is tuned to higher frequencies. It is also easier to build tunable oscillators than tunable electronic filters. In an analog television set, for example, if the superheterodyne design was not used, all the complicated filters that separate out the luminance carrier, chroma carrier and the audio subcarrier from the television signal would have to be made adjustable, and retuned each time the channel was changed.

- The total amplification of the receiver is divided between three amplifiers at different frequencies; the RF, IF, and audio amplifier. This reduces problems with feedback and parasitic oscillations that are encountered in receivers where most of the amplifier stages operate at the same frequency, as in the TRF receiver.

- The most important advantage is that better selectivity can be achieved by doing the filtering at the lower intermediate frequency. One of the most important parameters of a receiver is its bandwidth, the band of frequencies it accepts. In order to reject nearby interfering stations or noise, a narrow bandwidth is required. In all known filtering techniques, the bandwidth of the filter increases in proportion with the frequency, so by performing the filtering at the lower IF, rather than the frequency of the original radio signal fRF, a narrower bandwidth can be achieved. Modern FM and television broadcasting, cellphones and other communications services, with their narrow channel widths, would be impossible without the superheterodyne.

A problem with superhet receivers is that when the incoming radio signal f_{RF} is mixed with the local oscillator signal f_{LO}, *two* new frequencies (heterodynes) are created, one at the sum $f_{RF} + f_{LO}$, and one at the difference $f_{RF} - f_{LO}$ of the frequencies. Because of this, without an input filter the receiver can receive incoming RF signals at two different frequencies, one above the LO frequency: $f_{RF1} = f_{LO} + f_{IF}$, and one below the LO frequency: $f_{RF2} = f_{LO} - f_{IF}$. The receiver can be designed to receive on either of these two frequencies; the former is called *low-side injection* because the LO signal frequency is below the received frequency, while the latter is *high-side injection*. Whichever is chosen, the other frequency is called the *image frequency*. If the receiver is receiving a signal on one of these frequencies, for example f_{RF1}, any other radio station or radio noise on the other frequency f_{RF2} will be received also, interfering with the desired signal. Therefore, the superheterodyne requires a bandpass filter on the input to reject this image frequency. This image rejection filter does not need great selectivity, but as the receiver is tuned to different frequencies it must "track" in tandem with the LO.

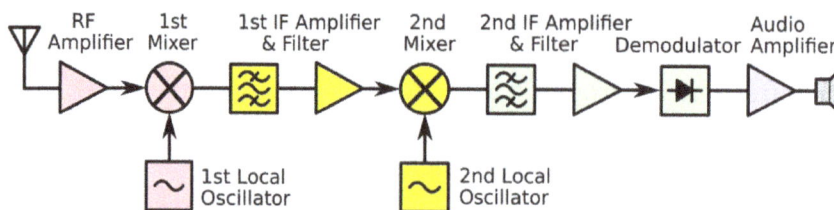

Double conversion superheterodyne block diagram

Double-conversion superheterodyne: In choosing the intermediate frequency (IF) used in a superhet, there is a tradeoff between image rejection and selectivity. Since the separation between the desired frequency and the image is $2f_{IF}$, the higher the IF, the farther apart these two frequencies are, and the better the image rejection. However, the lower the IF, the narrower the bandwidth the IF filter can achieve, and the better the selectivity. In order to achieve both good image rejection and selectivity, many modern superhet receivers use *two* intermediate frequencies; this is called a *dual-conversion* or *double-conversion* superheterodyne. The incoming RF signal is first mixed with one local oscillator signal in the first mixer to convert it to a high IF frequency, to allow efficient filtering out of the image frequency, then this first IF is mixed with a second local oscillator signal in a second mixer to convert it to a low IF frequency for good bandpass filtering. Some receivers even use triple-conversion.

Terminology used with superheterodyne receivers:

- *RF front end* - refers to all the components of the receiver up to and including the mixer; all the parts that process the signal at the original incoming radio frequency.

- *Converter* - in many superhet circuits the same tube or transistor functions as both the local oscillator and the mixer. This is called a converter.

- *IF strip* - The IF filter and amplifier usually consists of several tuned amplifier stages as in the TRF receiver. This is called the IF strip.

- *First detector, second detector* - the mixer tube or transistor is sometimes called the "first detector", while the demodulator that extracts the modulation from the IF signal is called the "second detector". In a dual-conversion superhet there are two mixers, so the demodulator is called the "third detector".

To receive radiotelegraph (CW) or single-sideband modulation (SSB), a beat frequency oscillator (BFO) is used.

Semiconductor Era

A modern smartphone has several digital radio transmitters and receivers to connect to different devices; a cellular receiver, a wireless modem, a bluetooth modem, and a GPS receiver

The invention of the transistor in 1947 revolutionized radio technology, making truly portable receivers possible, beginning with transistor radios in the late 1950s. Although portable vacuum tube radios were made, tubes were bulky and inefficient, consuming large amounts of power and requiring several large batteries to produce the filament and plate voltage. Transistors did not require a heated filament, reducing power consumption, and were smaller and much less fragile than vacuum tubes.

The development of integrated circuits (ICs) in the 1970s created another revolution, allowing an entire radio receiver to be put on a chip. ICs reversed the economics of vacuum tube radios; since the marginal cost of adding additional amplifying devices (transistors) to the chip was essentially zero, the size and cost of the receiver was dependent not on the active components as with vacuum tubes, but on the passive components; inductors and capacitors, which could not be integrated easily on the chip. As a result, the current trend in receivers is to use digital circuitry on the chip to do functions that were formerly done by analog circuits which require passive components. In a digital receiver the IF signal is sampled and digitized, and the bandpass filtering and detection functions are performed by digital signal processing (DSP) on the chip. Another benefit of DSP is that the properties of the receiver; channel frequency, bandwidth, gain, etc. can be dynamically changed by software to react to changes in the environment; these systems are known as software-defined radios or cognitive radio.

In modern wireless technology, radio transmitters and receivers are embedded in portable digital devices such as laptops, cell phones, GPS receivers, Bluetooth headsets, wireless routers and work automatically, in the background, to keep the device in touch with other devices, without wires. They transmit binary digital data by modulation methods such as frequency shift keying (FSK) allowing portable digital devices to communicate via wireless networks.

Digital Technologies

Many of the functions performed by analogue electronics can be performed by software instead. The benefit is that software is not affected by temperature, physical variables, electronic noise and manufacturing defects. For really high-performance receivers, such as satellite communications receivers and military/naval receivers, two-stage ("double conversion") and even three-stage ("triple conversion") superheterodyne processing is frequently used. Single-conversion receivers are rather simple-minded in their nature.

DSP Technology

DSP technology, short for digital signal processing, is the use of digital means to process signals and is coming into wide use in modern shortwave receivers. It is the basis of many areas of modern technology including cell phones, CD players, video recorders and computers. A digital signal is essentially a stream or sequence of numbers that relay a message through some sort of medium such as a wire. The primary benefit of DSP hardware in shortwave receivers is the ability to tailor the bandwidth of the receiver to current reception conditions and to the type of signal being listened to. A typical analog only receiver may have a limited number of fixed bandwidths, or only one, but a DSP receiver may have 40 or more individually selectable filters.

PC Controlled Radio Receivers

"PC radios", or radios that are designed to be controlled by a standard PC are controlled by specialized PC software using a serial port connected to the radio. A "PC radio" may not have a front-panel at all, and may be designed exclusively for computer control, which reduces cost.

Some PC radios have the great advantage of being field upgradable by the owner. New versions of the DSP firmware can be downloaded from the manufacturer's web site and uploaded into the flash memory of the radio. The manufacturer can then in effect add new features to the radio over time, such as adding new filters, DSP noise reduction, or simply to correct bugs.

A full-featured radio control program allows for scanning and a host of other functions and, in particular, integration of databases in real-time, like a "TV-Guide" type capability. This is particularly helpful in locating all transmissions on all frequencies of a particular broadcaster, at any given time. Some control software designers have even integrated Google Earth to the shortwave databases, so it is possible to "fly" to a given transmitter site location with a click of a mouse. In many cases the user is able to see the transmitting antennas where the signal is originating from.

Radio Control Software

The field of software control of PC radios has grown rapidly in the last several years, with developers making a number of advances. Since the Graphical User Interface or GUI interface PC to the radio has unlimited flexibility, any number of new features can be added by the software designer. Features that can be found in advanced control software programs today include a band table, GUI controls corresponding to traditional radio controls, local time clock and a UTC clock, signal strength meter, an ILG database for shortwave listening with lookup capability, scanning capability, text-to-speech interface, and integrated Conference Server.

Software-defined Radios

The next level in radio / software integration are so-called pure "software defined radios". The distinction here is that all filtering, modulation and signal manipulation is done in software, usually by a PC soundcard or by a dedicated piece of DSP hardware. There may be a minimal RF front-end or traditional radio that supplies an IF to the SDR. SDR's can go far beyond the usual demodulation capability of typical, and even high-end DSP shortwave radios. They can for example, record large swaths of the radio spectrum to a hard drive for "playback" at a later date. The same SDR that one minute is demodulating a simple AM broadcast may also be able to decode an HDTV broadcast in the next. A well known open-source project called GNU Radio is dedicated to evolving a high-performance SDR. All the source code for this SDR is freely downloadable and modifiable by anyone.

References

- Maxwell, W. M. W2DU (1990). Reflections: Transmission lines and antennas, 1st ed. Newington, CT: American Radio Relay League. ISBN 0-87259-299-5.

- H. Ward Silver, ed. ARRL Antenna Book. Newington, Connecticut: American Radio Relay League (2011), p. 24-22 ISBN 978-0-87259-694-8

- Joel R. Hallas. (2010). The ARRL Guide to Antenna Tuners, pg. 4-3. Newington, Connecticut: American Radio Relay League. ISBN 978-0-87259-098-4.

- Silver, H.W. (2014). The ARRL Handbook, 2015 Ed., pg. 20-16. Newington, CT: American Radio Relay League. ISBN 978-1-62595-019-2.

- Joel R. Hallas. (2010). The ARRL Guide to Antenna Tuners, pg. 7-4. Newington, Connecticut: American Radio Relay League, ISBN 978-0-87259-098-4

- Hall, Jerry (Ed.). (1988). ARRL Antenna Book, p. 25–18ff. Newington, CT: American Radio Relay League. ISBN 978-0-87259-206-3

- Lee, Thomas H. (2004). The Design of CMOS Radio-Frequency Integrated Circuits, 2nd Ed. UK: Cambridge University Press. pp. 1–8. ISBN 0521835399.

- Rudersdorfer, Ralf (2013). Radio Receiver Technology: Principles, Architectures and Applications. John Wiley and Sons. pp. 1–2. ISBN 111864784X.

- Nahin, Paul J. (2001). The Science of Radio: With Matlab and Electronics Workbench Demonstration, 2nd Ed. Springer Science & Business Media. pp. 45–48. ISBN 0387951504.

- Maver, William Jr. (August 1904). "Wireless Telegraphy To-Day". American Monthly Review of Reviews. New York: The Review of Reviews Co. 30 (2): 192. Retrieved January 2, 2016.

- Secor, H. Winfield (January 1917). "Radio Detector Development". Electrical Experimenter. New York: Experimenter Publishing Co. 4 (9): 652–656. Retrieved January 3, 2016.

- Hogan, John V. L. (April 1921). "The Heterodyne Receiver". The Electric Journal. Pittsburgh, USA: The Electric Journal. 18 (4): 116–119. Retrieved January 28, 2016.

- Haan, E. R. (December 1925). "Radio batteries, their care and selection". Popular Mechanics. New York: Popular Mechanics Co. 44 (6): 1010–1013. ISSN 0032-4558. Retrieved January 1, 2016.

- Langmuir, Irving (September 1915). "The Pure Electron Discharge and its Applications in Radio Telegraphy and Telephony" (PDF). Proceedings of the IRE. New York: Institute of Radio Engineers. 3 (3): 261–293. doi:10.1109/jrproc.1915.216680. Retrieved January 12, 2016.

- Tyne, Gerald F. J. (December 1943). "The Saga of the Vacuum Tube, Part 9" (PDF). Radio News. Chicago: Ziff-Davis. 30 (6): 30–31, 56, 58. Retrieved June 17, 2016.

- Armstrong, Edwin H. (April 1921). "The Regenerative Circuit". The Electrical Journal. Pittsburgh, PA: Westinghouse Co. 18 (4): 153–154. Retrieved January 11, 2016.

- In the early 1920s Armstrong, David Sarnoff head of RCA, and other radio pioneers testified before the US Congress on the need for legislation against radiating regenerative receivers. Wing, Willis K. (October 1924). "The Case Against the Radiating Receiver" (PDF). Broadcast Radio. New York: Doubleday, Page and Co. 5 (6): 478–482. Retrieved January 16, 2016.

- Armstrong, Edwin H. (February 1921). "A new system of radio frequency amplification". Proceedings of the Inst. of Radio Engineers. New York: Institute of Radio Engineers. 9 (1): 3–11. Retrieved December 23, 2015.

- Crookes, William (February 1, 1892). "Some Possibilities of Electricity". The Fortnightly Review. London: Chapman and Hall. 51: 174–176. Retrieved August 19, 2015.

- Hazeltine, Louis A. (March 1923). "Tuned Radio Frequency Amplification With Neutralization of Capacity Coupling" (PDF). Proc. of the Radio Club of America. New York: Radio Club of America. 2 (8): 7–12. Retrieved March 7, 2014.

3

Types of Antenna

A vital part of the digital wireless industry is the antenna. There are several different types of antennas; some of them which have been mentioned in the chapter such as monopole antenna, loop antenna, random wire antenna, smart antenna, etc.

Monopole Antenna

Mast radiator monopole antenna used for broadcasting. AM radio station WARE, Warren, Massachusetts, USA.

A monopole antenna is a class of radio antenna consisting of a straight rod-shaped conductor, often mounted perpendicularly over some type of conductive surface, called a *ground plane*. The driving signal from the transmitter is applied, or for receiving antennas the output signal to the receiver is taken, between the lower end of the monopole and the ground plane. One side of the antenna feedline is attached to the lower end of the monopole, and the other side is attached to the ground plane, which is often the Earth. This contrasts with a dipole antenna which consists of two identical rod conductors, with the signal from the transmitter applied between the two halves of the antenna.

The monopole is a resonant antenna; the rod functions as an open resonator for radio waves, oscillating with standing waves of voltage and current along its length. Therefore, the length of the antenna is determined by the wavelength of the radio waves it is used with. The most common or fundamental form is the *quarter-wave monopole*, in which the antenna is approximately ¼ of the wavelength of the radio waves. The monopole antenna was invented in 1895 by radio pioneer Guglielmo Marconi; for this reason it is sometimes called the *Marconi antenna*. Common types of monopole antenna are the whip, rubber ducky, helical, random wire, umbrella, inverted-L and T-antenna, inverted-F, mast radiator, and ground plane antennas.

History

The monopole antenna was invented in 1895 by radio pioneer Guglielmo Marconi during his historic first experiments in radio communication. He began by using Hertzian dipole antennas consisting of two identical horizontal wires ending in metal plates. He found by experiment that if instead of the dipole, one side of the transmitter was connected to a wire suspended overhead, and the other side was connected to the Earth, he could transmit for longer distances. For this reason the monopole is also called a *Marconi antenna*, although Alexander Popov independently invented it at about the same time.

Radiation Pattern

Like a dipole antenna, a monopole has an omnidirectional radiation pattern. That is it radiates equal power in all azimuthal directions perpendicular to the antenna, but the radiated power varies with elevation angle, with the radiation dropping off to zero at the zenith, on the antenna axis. It radiates vertically polarized radio waves.

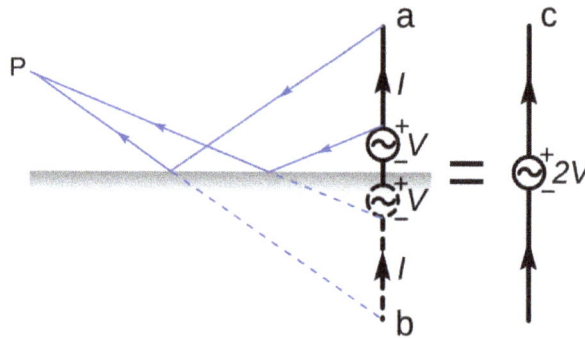

Showing the monopole antenna has the same radiation pattern over perfect ground as a dipole in free space with twice the voltage

A monopole can be visualized *(right)* as being formed by replacing the bottom half of a vertical dipole antenna *(c)* with a conducting plane (ground plane) at right-angles to the remaining half. If the ground plane is large enough, the radio waves from the remaining upper half of the dipole *(a)* reflected from the ground plane will seem to come from an image antenna *(b)* forming the missing half of the dipole, which adds to the direct radiation to form a dipole radiation pattern. So the pattern of a monopole with a perfectly conducting, infinite ground plane is identical to the top half of a dipole pattern, with its maximum radiation in the horizontal direction, perpendicular to the antenna. Because it radiates only into the space above the ground plane, or half the space of a dipole antenna, a monopole antenna will have a gain of twice (3 dB greater than) the gain of a similar dipole antenna, and a radiation resistance half that of a dipole. Since a half-wave dipole has a gain of 2.19 dBi and a radiation resistance of 73 ohms, a quarter-wave monopole, the most common type, will have a gain of 2.19 + 3 = 5.19 dBi and a radiation resistance of about 36.8 ohms if it is mounted above a good ground plane.

The general effect of electrically small ground planes, as well as imperfectly conducting earth grounds, is to tilt the direction of maximum radiation up to higher elevation angles.

Types

The ground plane used with a monopole may be the actual earth; in this case the antenna is mounted on the ground and one side of the feedline is connected to an earth ground at the base of the antenna. This design is used for the mast radiator antennas employed in radio broadcasting at low frequencies, as well as other low frequency antennas such as the T-antenna and umbrella antenna. At VHF and UHF frequencies the size of the ground plane needed is smaller, so artificial ground planes are used to allow the antenna to be mounted above the ground. A common type of monopole antenna at these frequencies consists of a quarter-wave whip antenna with a ground plane consisting of several wires or rods radiating horizontally or diagonally from its base; this is called a ground-plane antenna. At gigahertz frequencies the metal surface of a car roof or airplane body makes a good ground plane, so car cell phone antennas consist of short whips mounted on the roof, and aircraft communication antennas frequently consist of a short conductor in an aerodynamic fairing projecting from the fuselage; this is called a *blade antenna*. The most common antenna used in mobile phones is the inverted-F antenna, which is a variant of the inverted-L monopole. Bending over the antenna saves space and keeps the it within the bounds of the mobile's case but the antenna then has a very low impedance. To improve the match the antenna is not fed from the end, rather some intermediate point, and the end is grounded instead. The quarter-wave whip and "Rubber Ducky" antennas used with handheld radios such as walkie-talkies and cell phones are also monopole antennas. These don't use a ground plane, and the ground side of the transmitter is just connected to the ground connection on its circuit board. The hand and body of the person holding them may function as a rudimentary ground plane.

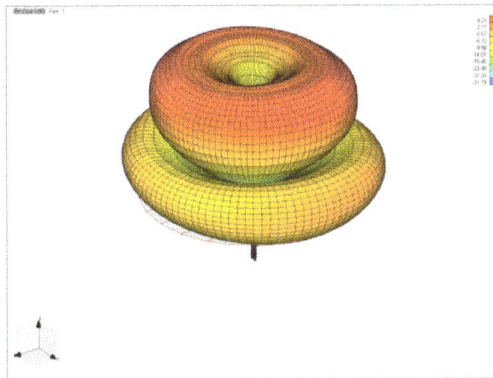

Radiation pattern of 3/2 wavelength monopole. Monopole antennas up to 1/4 wavelength long have a single "lobe", with field strength declining monotonically from a maximum in the horizontal direction, but longer monopoles have more complicated patterns with several conical "lobes" (radiation maxima) directed at angles into the sky.

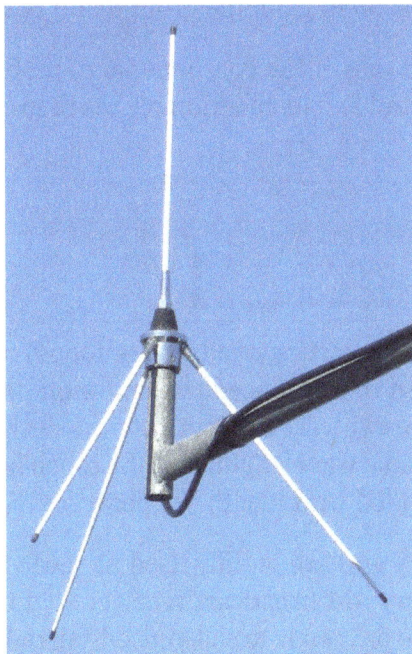

VHF ground plane antenna, a type of monopole antenna used at high frequencies. The three conductors projecting downward are the ground plane

Monopole Broadcasting Antennas

When used for radio broadcasting, the radio frequency power from the broadcasting transmitter is fed across the base insulator between the tower and a ground system. The ideal ground system for AM broadcasters comprises at least 120 buried copper or phosphor bronze radial wires at least one-quarter wavelength long and a ground-screen in the immediate vicinity of the tower. All the ground system components are bonded together, usually by welding, brazing or using coin silver solder to help reduce corrosion. Monopole antennas that use guy-wires for support are called masts in some countries. In the United States, the term "mast" is generally used to describe a pipe supporting a smaller antenna, so both self-supporting and guy-wire supported radio antennas are simply called monopoles if they stand alone. If multiple monopole antennas are used in order to control the direction of radio frequency (RF) propagation, they are called directional antenna arrays.

In the United States, the Federal Communications Commission (FCC) requires that the transmitter power input to the antenna be measured and maintained. The power input is calculated as the square of the measured current, i, flowing into the antenna from the transmission line multiplied by the real part of the antenna's feed-point impedance, r.

$$P = i^2 r$$

This impedance is periodically measured to verify the stability of the antenna and ground system. Normally, an impedance matching network matches the impedance of the antenna to the impedance of the transmission line feeding it.

Examples of monopole antennas are:

- the whip antenna
- the mast radiator (when isolated from the ground and bottom-fed)

Monopole antennas have become one of the components of mobile and internet networks across the globe. Their relative low cost and fast installation makes them an obvious choice for developing countries.

Loop Antenna

A loop antenna is a radio antenna consisting of a loop (or loops) of wire, tubing, or other electrical conductor with its ends connected to a balanced transmission line (possibly via a balun). Within this physical description there are two very distinct antenna designs: the small loop (or *magnetic loop*) with a size much smaller than a wavelength, and the much larger resonant loop antenna with a circumference close to the intended wavelength of operation.

Small loops have a poor efficiency and are mainly used as receiving antennas at low frequencies. Except for car radios, almost every AM broadcast receiver sold has such an antenna built inside it or directly attached to it. These antennas are also used for radio direction finding. In amateur radio, loop antennas are often used for low profile operating where larger antennas would be inconvenient, unsightly, or banned. Loop antennas are relatively easy to build.

A small loop antenna, also known as a magnetic loop, generally has a circumference of less than one tenth of a wavelength, in which case there will be a relatively constant current distribution along the conductor. As the frequency or the size is increased, a standing wave starts to develop in the current, and the antenna starts to acquire some of the characteristics of a resonant loop (but isn't resonant); these intermediate cases thus cannot be analyzed using the concepts developed for the small and resonant loop antennas described below.

Resonant loop antennas are relatively large, governed by the intended wavelength of operation. Thus they are typically used at higher frequencies, especially VHF and UHF, where their size is manageable. They can be viewed as a folded dipole deformed into a different shape, and have rather similar characteristics such as a high radiation efficiency.

Similar and Dissimilar Devices

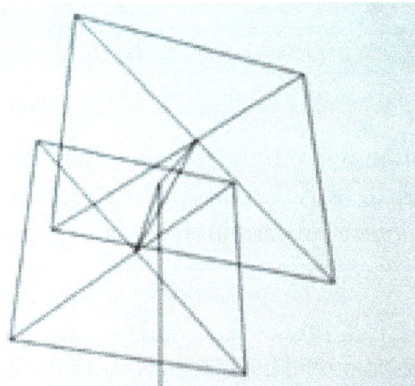

The "Quad antenna" is a resonant loop in a square shape; this one also includes a parasitic element

Although a resonant loop may be in the shape of a circle, distorting it into a somewhat different closed shape does not greatly alter its characteristics. For instance, the quad antenna popular in amateur radio consists of a resonant loop (and usually additional parasitic elements) in a rectangular shape, so that it can be constructed of wire strung across a supporting 'X' frame. Or a large loop can be completely collapsed into a line, in which case it is termed a folded dipole. In the case of large loops, such as the quad or a folded dipole, the antenna's resonant frequency is determined by the circumference of the loop. On the other hand a small loop antenna is used for wavelengths much bigger than the loop itself; its radiation resistance and efficiency are instead dependent on the area enclosed by the loop (and number of turns). For a given loop area, the length of the conductor (and thus its net loss resistance) is minimized if the shape is a circle, making that case optimum for small loops.

Although it has a superficially similar appearance, the so-called halo antenna is not technically a loop since it possesses a break in the conductor opposite the feed point. Its characteristics are unlike that of either sort of loop antenna here described.

Also outside the scope of this article is the use of coupling coils for inductive (magnetic) transmission systems including LF and HF (rather than UHF) RFID tags and readers. Although these do use radio frequencies, and involve the use of small loops (loosely described as "antennas" in the trade) which may be physically indistinguishable from the small loop antennas discussed here, such systems are not designed to transmit radio waves (electromagnetic waves). They are near field systems involving alternating magnetic fields only, and may be analyzed as poorly coupled transformer windings; their performance criteria are dissimilar to radio antennas as discussed here.

Resonant Loop Antennas

A loop antenna for amateur radio under construction

The large or resonant loop antenna can be seen as a folded dipole which has been reformed into a circle (or square, etc.). In order to be resonant (having a purely resistive driving point impedance) the loop requires a circumference approximately equal to one wavelength (however it will also be resonant at odd multiples of a wavelength).

Contrary to the small loop antenna, this design radiates in the direction *normal* to the plane of the loop (thus in two opposite directions). Therefore these loops are normally installed with the plane of the loop in the vertical direction, and may be rotatable. Compared to a dipole or folded dipole, it then transmits less toward the sky or ground, giving it a slightly higher gain (about 10% higher) in the horizontal direction. Further directionality can be obtained by using a loop whose circumference is not one but 3 or 5 wavelengths. However it is more common to increase gain using an array of driven loops or a Yagi configuration including parasitic loop elements. The latter is widely used in amateur radio where it is referred to as a quad antenna, with the loops being square as they are usually constructed with wires held taut in between the rigid "X" structures.

The polarization of such an antenna is not obvious by looking at the loop itself, but depends on the feed point (where the transmission line is connected). If a vertically oriented loop is fed at the bottom it will be horizontally polarized; feeding it from the side will make it vertically polarized.

Small Loops

Small loop antennas are much less than a wavelength in size, and are mainly (but not always) used as receiving antennas at lower frequencies.

Magnetic Vs. Electrical Antennas

The small loop antenna is also known as a *magnetic loop* since it behaves electrically as a coil (inductor) with a limited but non-negligible radiation resistance due to its small size compared to one wavelength. It can be analyzed as coupling directly to the magnetic field in the region near the antenna – the opposite of the principle of a Hertzian dipole which couples directly to the electric field. The magnetic field then gives rise to an electric field through Faraday's law of induction and a full blown electromagnetic wave in the region far from the antenna.

Although a full 2.7 meters in diameter, this receiving antenna is a "small" loop for LF/MF wavelengths.

Radiation Pattern and Polarization

Surprisingly, the radiation and receiving pattern of a small loop is quite opposite that of a large loop (whose circumference is close to one wavelength). Since the loop is much smaller than a wavelength, the current at any one moment is nearly constant round the circumference. By symmetry it can be seen that the voltages induced along the sides of the loop will cancel each other when a signal arrives along the loop axis. Therefore there is a null in that direction. Instead, the radiation pattern peaks in directions lying in the plane of the loop, because signals received from sources in that plane do not quite cancel owing to the phase difference between the arrival of the wave at the near side and far side of the loop. Increasing that phase difference by increasing the size of the loop has a large impact in increasing the radiation resistance and the resulting antenna efficiency.

Another way of looking at a small loop as an antenna is to consider it simply as an inductive coil coupling to the magnetic field in the direction *perpendicular* to plane of the coil, according to Ampère's law. Then consider a propagating radio wave also perpendicular to that plane. Since the magnetic (and electric) fields of an electromagnetic wave in free space are transverse (no component in the direction of propagation), it can be seen that this magnetic field and that of a small loop antenna will be at right angles, and thus not coupled. For the same reason, an electromagnetic wave propagating through the plane of the loop, with its magnetic field perpendicular to that plane, *is* coupled to the magnetic field of the coil. Since the transverse magnetic and electric fields of a propagating electromagnetic wave are at right angles, the electric field of such a wave is also in the plane of the loop, and thus the antenna's *polarization* (which is always specified as being the orientation of the electric, not the magnetic field) is said to be in that plane.

Thus mounting the loop in a horizontal plane will produce an omnidirectional antenna which is horizontally polarized; mounting the loop vertically yields a weakly directional antenna with vertical polarization.

Small Loop Receiving Antenna

Loopstick antenna from an AM radio having two windings, one for long wave and one for medium wave (AM broadcast) reception. Typically 10 cm long, these loop antennas are usually hidden inside the radio receiver.

AM broadcast radios (and other consumer low frequency receivers) typically use small loop antennas; a variable capacitor connected across the loop forms a tuned circuit that also tunes the receivers input stage as that capacitor tracks the main tuning. A multiband receiver may contain tap points along the loop winding in order to tune the loop antenna at widely different frequencies. In older (and physically larger) AM radios, the small loop might consist of dozens of turns of wire in a loop on the back side of the radio (a *frame antenna*). In modern radios, the loop antenna often is wound on a ferrite rod; the ferrite rod allows a physically small antenna to have a larger effective antenna area. The resulting coil and core is called a *loopstick antenna*, a *ferrite rod antenna*, a *ferrite rod aerial*, a *Ferroceptor*, a *ferrod antenna*, or a *ferrite antenna*. The term *loopstick* refers to the underlying loop antenna and the stick shape of the ferrite rod.

Small loop antennas are lossy and inefficient, but they can make practical receiving antennas in the medium-wave (520–1610 kHz) band and below, where the antenna inefficiency is masked by large amounts of atmospheric noise. Loop antennas are often wound with litz wire to reduce skin effect losses.

Antenna Efficiency

Since a small loop antenna is essentially a coil, its electrical impedance is inductive, with an inductive reactance much greater than its radiation resistance. In order to couple to a transmitter or receiver, the inductive reactance is normally canceled with a parallel capacitance. Since a good loop antenna will have a high Q factor, this capacitor must be variable and is adjusted along with the receiver's tuning.

The radiation resistance R_R of a small loop is generally much smaller than the loss resistance R_L due to the conductors comprising the loop, leading to a poor antenna efficiency. Consequently, most of the transmitted or received power will be dissipated as heat.

So much wasted signal power is a disaster for a transmitting antenna, however in a receiving antenna the inefficiency is not important at frequencies below about 10 MHz . At those lower frequencies atmospheric noise (static) and man-made noise (interferance) dominate over the noise generated inside the receiver itself (*thermal* or *Johnson noise*). Any increase in signal strength increases both the signal and the external noise in equal proportion, leaving the signal-to-noise ratio unchanged. (CCIR 258; CCIR 322.)

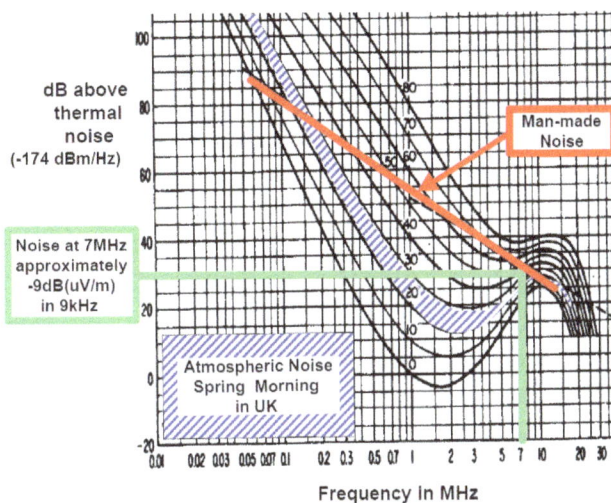

Amount of atmospheric noise for LF, MF, and HF spectrum according CCIR 322

For example, at 1 MHz the man-made noise might be 55 dB above the thermal noise floor. If a small loop antenna's loss is 50 dB (as if the antenna included a 50 dB attenuator) the electrical inefficiency of that antenna will have little influence on the receiving system's signal-to-noise ratio. In contrast, at quieter frequencies above about 20 MHz an antenna with a 50 dB loss could degrade the received signal-to-noise ratio by up to 50 dB, resulting in terrible performance. Copper losses are often minimized by the use of spiderweb or basket winding construction and Litz wire.

Insensitivity to Locally Generated Interference

Due to its direct coupling to the magnetic field, unlike most other antenna types, the small loop is relatively insensitive to electric-field noise from nearby sources. No matter how close the electrical interference is to the loop, its effect will not be much greater than if it were a quarter wavelength away. This is valuable inasmuch as most sources of interference with radio frequency content, such as sparking at commutators or corona discharge, produce electric, not magnetic fields.

Small loops are especially used in the AM broadcast band and generally at lower frequencies where resonant antennas are of an impractical size. At those frequencies the near-field is physically quite large. This provides a considerable advantage for loop antennas which are relatively insensitive to the main interference sources encountered.

The same principle makes a small loop particularly sensitive to sources of *magnetic* noise in its near field. Likewise, a Hertzian (short) dipole couples directly with the electric field and is relatively immune to locally produced magnetic noise. However at radio frequencies nearby sources of magnetic interference are generally not an issue. In either case the small antenna's immunity does not extend to noise sources outside of the near field: Noise sources over one wavelength distant, whether originating as an electric or magnetic field, are received simply as electromagnetic waves. Noise from outside any antenna's near field will be received equally well by any antenna sensitive to a radio transmitter from the direction of that noise source.

Receiver Input Tuning

Small loop receiving antennas are also almost always resonated using a parallel capacitor, which makes their reception narrow-band, sensitive only to a very specific frequency. This allows the antenna, in conjunction with a (variable) tuning capacitor, to act as a tuned input stage to the receiver's front-end, in lieu of a coil.

Small Loops as Transmitting Antennas

Due to their small radiation resistance and consequent electrical inefficiency, small loops are seldom used as transmitting antennas, where one is trying to couple most of the transmitter's power to the electromagnetic field. Nevertheless small loops are sometimes used in applications in which a resonant antenna (with elements around a quarter of a wavelength in size) would simply be too large to be practical. Since *any* antenna much smaller than a wavelength suffers from inefficiency, a loop might not be the worst choice for medium wave and lower frequencies. Small loops (typically 18 to 39 inches in diameter when used from 29.7-7 MHz) are becoming popular as transmitting (as well as receiving) antennas.

The radiation efficiency is greatly boosted by making the outer loop larger (compared to one only used for receiving) insofar as that is possible in a given application, with circumferences ideally greater than $1/_{10}$ of a wavelength. Note that the increased size of the now not-so-small loop alters its radiation pattern, as the assumption of currents being totally in phase along the circumference of the loop begins to break down. In addition to making the geometric loop larger, efficiency is also increased by using larger conductors in order to reduce the loss resistance, and plating the outer conductor surfaces with silver or non-anodized aluminum.

Small loops are used in land-mobile radio (mostly military) at frequencies between 3–7 MHz, because of their ability to direct energy upwards, unlike a conventional whip antenna. This enables Near Vertical Incidence Skywave (NVIS) communication up to 300 km in mountainous regions. In this case a typical radiation efficiency of around 1% is acceptable because signal paths can be established with 1 Watt of radiated power or less when a transmitter generating 100 Watts is used. In military use, the antenna elements can be 2-3 inches in diameter.

One practical issue with small loops as transmitting antennas is that the loop not only has a very large current going through it, but also has a very high voltage on its terminals, typically kilo-Volts when fed with only a few Watts of transmitter power. This requires a rather expensive and physically large resonating capacitor with a large breakdown voltage, in addition to having minimal dielectric loss (normally requiring an air-gap capacitor).

To keep a balanced point of view, it is important to note that a vertical or dipole antenna that is short compared to a wavelength, matched using a small loading coil *also* has a high voltage present at the loading coil. The difference being that since the loop antenna is already physically large in order to reduce loss and carry high current, high voltage breakdown is not usually as much of an issue.

As for any antenna system, efficient electrical coupling requires impedance matching. For a small loop tuned with a parallel capacitor the resulting large (resistive) impedance (or for a small loop tuned with a series capacitor the resulting small impedance) will not be a good match to a standard transmission line or transmitter. In addition to other common impedance matching techniques, this is sometimes accomplished by connecting the transmission line not directly to the loop but to a smaller *feed loop*, typically $\frac{1}{8}$ to $\frac{1}{5}$ the size of the loop antenna. Essentially acting as a step-up transformer, power is inductively coupled from the feed loop to the main loop which itself is connected to the resonating capacitor and is responsible for radiating most of the power.

Direction Finding with Loops

Since the directional response of small loop antennas includes a sharp null in the direction normal to the plane of the loop, they are used in radio direction finding at longer wavelengths.

The procedure is to rotate the loop antenna to find the direction where the signal vanishes – the "null" direction. Since the null occurs at two opposite directions along the axis of the loop, other means must be employed to determine which side of the antenna the "nulled" signal is on. One method is to rely on a second loop antenna located at a second location, or to move the receiver to that other location, thus relying on triangulation.

Instead of triangulation, a second dipole or vertical antenna can be electrically combined with a loop or a loopstick antenna. Called a *sense antenna*, connecting the second antenna changes the combined radiation pattern to a cardioid, with a null in only one, less precise direction. The general direction of the transmitter can be determined using the sense antenna, and then disconnecting the sense antenna returns the sharp nulls in the loop antenna pattern, allowing a precise bearing to be determined.

Loop antenna, receiver, and accessories used in amateur radio direction finding at 80 meter wavelength (3.5 MHz).

Random Wire Antenna

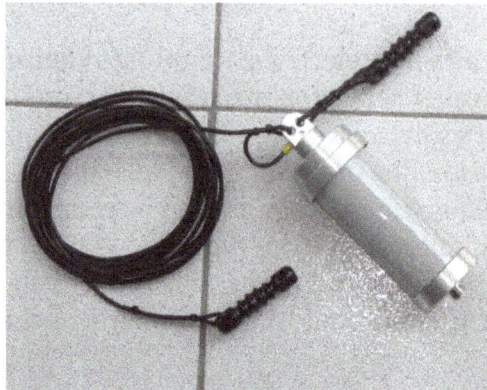

A wire antenna kit, with a coil of wire, strain insulators and a balun. When installed the wire is supported by buildings or trees using the insulators to prevent a short circuit to ground.

A random wire antenna is a radio antenna consisting of a long wire suspended above the ground, whose length does not bear a relation to the wavelength of the radio waves used, but is typically chosen more for convenience. The wire may be straight or it may be strung back and forth between trees or walls just to get enough wire into the air; this type of antenna sometimes is called a zig-zag antenna. Such antennas are usually not as effective as antennas whose length is adjusted to resonate at the wavelength to be used. Random wire antennas are a type of monopole antenna and the other side of the receiver or transmitter antenna terminal must be connected to an earth ground.

They are widely used as receiving antennas on the long wave, medium wave, and short wave bands, as well as transmitting antennas on these bands for small outdoor, temporary or emergency transmitting stations, as well as in situations where more permanent antennas cannot be installed.

Random Wire and Long Wire

Often random wire antennas are also referred to as long-wire antenna. Long-wire antennas require a length greater than a quarter-wavelength ($\lambda/4$) of the radio waves (most consider a true long wire to be least one wavelength), whereas random wire antennas have no such constraint.

Radiation Pattern

The radiation pattern of a straight random wire antenna is unpredictable and depends on its electrical length; its length measured in wavelengths (λ) of the radio waves used. The radiation will drop off to zero on the axis; however it may have several lobes (maxima) at angles to the antenna axis. Under about 0.6λ a wire antenna will have a single lobe with a maximum at right angles to the axis. Above this the lobe will split into two conical lobes with their maximum directed at equal angles to the wire, and a null between them. This results in four azimuth angles at which the gain is maximum. As the length of wire in wavelengths increases, the number of lobes increases and the maxima become increasingly sharp. A folded or zig-zag antenna will exhibit a pattern that is even more complicated and difficult to predict.

Long wire antennas are reported to be more effective for reception than multielement antennas such as Yagi or quad antennas with the same length of wire. Due to the length of the antenna, with multipath propagation there is a *diversity effect*; radio waves which interfere and cancel at one part of the antenna may not cancel at another part, resulting in more reliable overall reception.

Construction

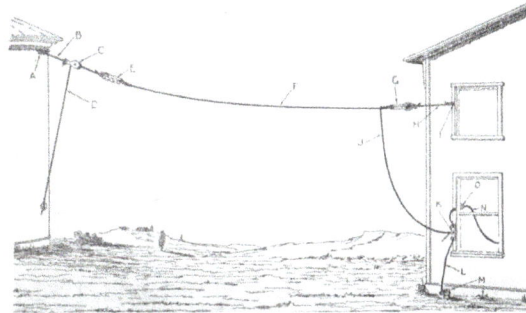

A typical permanent wire antenna strung between two buildings. This example has a *lightning switch* to ground the antenna for safety during electrical storms.

A random wire antenna usually consists of a long (at least one quarter wavelength) wire with one end connected to the radio and the other in free space, arranged in any way most convenient for the space available. Ideally, it is a straight wire strung as high as possible between trees or buildings, the ends insulated from supports with strain insulators. Typically it is made from number 12 or 14 AWG (1.6 to 2.0 mm (0.063 to 0.079 in) diameter) copperclad wire. Folding the wire into a zigzag pattern to fit in a limited space such as an apartment or attic will reduce effectiveness and make theoretical analysis extremely difficult. (The added length helps more than the folding typically hurts.)

If used for transmitting, a random wire antenna usually will also require an antenna tuner, as it has an unpredictable impedance that varies with frequency. One side of the output of the tuner is connected directly to the antenna, without a transmission line, the other to a good earth ground. A quarter-wavelength sized wire works best, and unless fed through a unun, a half-wavelength will exceed the matching ability of most tuners. Even without a good earth, the antenna will also radiate, but it will do so by coupling capacitively to any nearby conducting material; nevertheless this is not recommended. The ground for a random wire antenna may be chosen by experimentation. Grounds could be returned to a nearby cold water pipe or a wire approximately one-quarter wavelength long, or can be replaced by randomly laid-out quarter-wavelength counterpoise wires attached to the ground connection.

RF feedback, or 'backlash current', can be an issue. RF feedback can be minimized by selecting a wire length that causes the low feed-point impedance at a current loop to occur at the transmitter. Alternately, a remote tuner can be fed with feedline, and the tuner located on the antenna.

Beverage Antenna

The Beverage antenna or "wave antenna" is a long wire receiving antenna mainly used in the low frequency and medium frequency radio bands, invented by Harold H. Beverage in 1921. It is used by amateur radio, shortwave listening, and longwave radio DXers and military applications.

A Beverage antenna consists of a horizontal wire from one-half to several wavelengths long (hundreds of feet at HF to several kilometres for longwave) suspended above the ground, with the feedline to the receiver attached to one end and the other terminated through a resistor to ground. The antenna has a unidirectional radiation pattern with the main lobe of the pattern at a shallow angle into the sky off the resistor-terminated end, making it ideal for reception of long distance skywave (skip) transmissions from stations over the horizon which reflect off the ionosphere. However the antenna must be built so the wire points at the location of the transmitter. Some Beverage antennas use a two-wire design that allows reception in two directions from a single Beverage antenna. Other designs use sloped ends where the center of the antenna is six to eight feet high and both ends of the antenna gradually slope downwards towards the termination resistor and matching transformer.

The advantages of the Beverage are excellent directivity, and wider bandwidth than resonant antennas. It's disadvantages are its physical size, requiring considerable land area, and inability to rotate to change the direction of reception. Installations often use multiple antennas to provide wide azimuth coverage.

History

Harold H. Beverage experimented with receiving antennas similar to the Beverage antenna in 1919 at the Otter Cliffs Radio Station. He discovered in 1920 that an otherwise nearly bidirectional long wire antenna becomes uni-directional by placing it close to the lossy earth and by terminating one end of the wire with a resistor. By 1921, Beverage long wave receiving antennas up to nine miles (14 km) long had been installed at RCA's Riverhead, New York, Belfast, Maine, Belmar, New Jersey, and Chatham, Massachusetts receiver stations for transatlantic radiotelegraphy traffic. The antenna was patented in 1921 and named for its inventor Harold H. Beverage. Perhaps the largest Beverage antenna—an array of four phased Beverages three miles (5 km) long and two miles (3 km) wide—was built by AT&T in Houlton, Maine for the first transatlantic telephone system opened in 1927.

Description

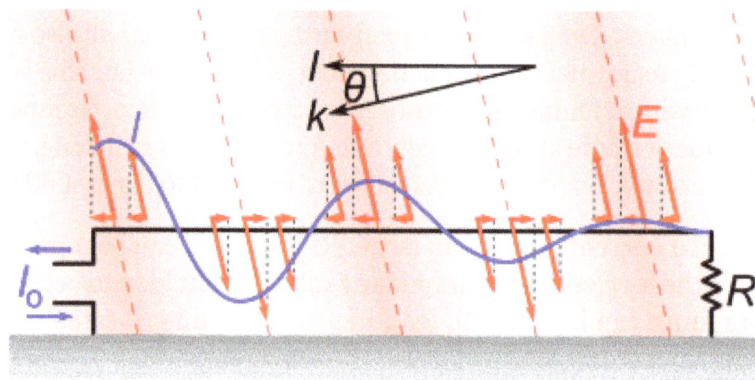

Animation showing how the antenna works. Due to ground resistance the electric field of the radio wave (E, big red arrows) is at an angle θ to the vertical, creating a horizontal component parallel to the antenna wire (small red arrows). The horizontal electric field creates a traveling wave of oscillating current (I, blue line) and voltage along the wire, which increases in amplitude with distance from the end. When it reaches the driven end (left), the current passes through the transmission line to the receiver. Radio waves in the other direction, toward the terminated end, create traveling waves which are absorbed by the terminating resistor R, so the antenna has a unidirectional pattern.

A Beverage antenna that can be improvised for military field communications, from a U.S. Army field manual. Rather than being grounded, the resistor is attached to a second lower wire which serves as a counterpoise, an artificial ground for the transmitter. The antenna's main lobe, its direction of greatest sensitivity, is to the right, off the end of the wire that is terminated in the resistor.

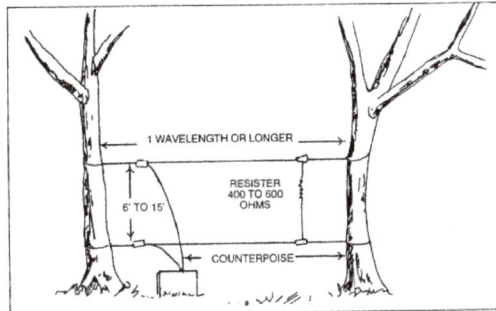

Figure D-11. Long-wire antenna.

The Beverage antenna consists of a horizontal wire one-half to several wavelengths long, suspended close to the ground, usually 10 to 20 feet high, pointed in the direction of the signal source. At the end toward the signal source it is terminated by a resistor to ground approximately equal in value to the characteristic impedance of the antenna considered as a transmission line, usually 400 to 800 ohms. At the other end it is connected to the receiver with a transmission line, through a balun to match the line to the antenna's characteristic impedance.

How it Works

Unlike other wire antennas such as dipole or monopole antennas which act as resonators, with the radio currents traveling in both directions along the element, bouncing back and forth between the ends as standing waves, the Beverage antenna is a traveling wave antenna; the radio frequency current travels in one direction along the wire, in the same direction as the radio waves. The lack of resonance gives it a wider bandwidth than resonant antennas. It receives vertically polarized radio waves, but unlike other vertically polarized antennas it is suspended close to the ground, and requires some resistance in the ground to work.

The Beverage antenna relies on "wave tilt" for its operation. At low and medium frequencies, a vertically polarized radio frequency electromagnetic wave traveling close to the surface of the earth with finite ground conductivity sustains a loss that causes the wavefront to "tilt over" at an angle. The electric field is not perpendicular to the ground but at an angle, producing an electric field component parallel to the Earth's surface. If a horizontal wire is suspended close to the Earth and approximately at a right angle to the wave front, the electric field generates an oscillating RF current wave traveling along the wire, propagating in the same direction as the wavefront. The RF currents traveling along the wire add in phase and amplitude throughout the length of the wire, producing maximum signal strength at the far end of the antenna where the receiver is connected.

The antenna wire and the ground under it together can be thought of as a "leaky" transmission line which absorbs energy from the radio waves. The velocity of the current waves in the antenna is less than the speed of light due to the ground. The velocity of the wavefront along the wire is also less than the speed of light due to its angle. At a certain angle θ_{max} the two velocities are equal. At this angle the gain of the antenna is maximum, so the radiation pattern has a main lobe at this angle. The angle of the main lobe is

$$\theta_{max} = \arccos\left(1 - \frac{\lambda}{2L}\right),$$

where

$\qquad$ L is the length of the antenna wire,

$\qquad$ λ is the wavelength.

The antenna has a unidirectional reception pattern, because RF signals arriving from the other direction, from the receiver end of the wire, induce currents propagating toward the terminated end, where they are absorbed by the terminating resistor.

The antenna is particularly suited to receive radio waves reflected from the ionosphere, called skywave, or "skip" propagation, which is used for long-distance communication at shortwave frequencies. The radio waves typically arrive at the Earth's surface with shallow angles (wave tilt) of approximately 5 to 45 degrees, which is a good match to the antenna's direction of maximal gain (main lobe).

Gain

While Beverage antennas have excellent directivity, because they are close to lossy Earth, they do not produce absolute gain; their gain is typically from −20 to −10 dBi. This is rarely a problem, because the antenna is used at frequencies where there are high levels of atmospheric radio noise. At these frequencies the atmospheric noise, and not receiver noise, determines the signal-to-noise ratio, so an inefficient antenna can be used. The antenna is not used as a transmitting antenna, since a large portion of the drive power is wasted in the terminating resistor. The Beverage antenna is a good receiving antenna because it offers excellent directivity over a broad bandwidth, albeit with relatively large size.

Directivity increases with the length of the antenna. While directivity begins to develop at a length of only 0.25 wavelength, directivity becomes more significant at one wavelength and improves steadily until the antenna reaches a length of about two wavelengths. In Beverages longer than two wavelengths, directivity does not increase because the currents in the antenna cannot remain in phase with the radio wave.

The Beverage antenna is most frequently deployed as a single wire. A dual-wire variant is sometimes utilized for rearward null steering or for bidirectional switching. The antenna can also be implemented as an array of 2 to 128 or more elements in broadside, endfire, and staggered configurations, offering significantly improved directivity otherwise very difficult to attain at these frequencies. A four-element broadside/staggered Beverage array was used by AT&T at their longwave telephone receiver site in Houlton, Maine. Very large phased Beverage arrays of 64 elements or more have been implemented for receiving antennas for over-the-horizon radar systems.

Implementation

A single-wire Beverage antenna is typically a single straight copper wire, between one and two wavelengths long, running parallel to the Earth's surface in the direction of the desired signal. The wire is suspended by insulated supports above the ground. A non-inductive resistor approximately equal to the characteristic impedance of the wire, about 400 to 600 ohms, is connected from the far end of the wire to a ground rod. The other end of the wire is connected to the feedline to the receiver.

The driving impedance of the antenna is equal to the characteristic impedance of the wire with respect to ground, somewhere between 400 and 800 ohms, depending on the height of the wire. Typically a length of 50-ohm or 75-ohm coaxial cable would be used for connecting the receiver to the antenna endpoint. A matching transformer should be inserted between any such low-impedance transmission line and the higher 470-ohm impedance of the antenna. A transformer with a turns ratio of 3:1 would provide an impedance transformation of 9:1, which will match the antenna to a 50-ohm transmission line. A transformer with a turns ratio of 5:2 would provide an impedance transformation of 6.25:1, which will match the antenna to a 75-ohm transmission line. Alternatively, a parallel-wire transmission line (twin lead) of 600 ohms makes a fairly good match to the antenna.

Rhombic Antenna

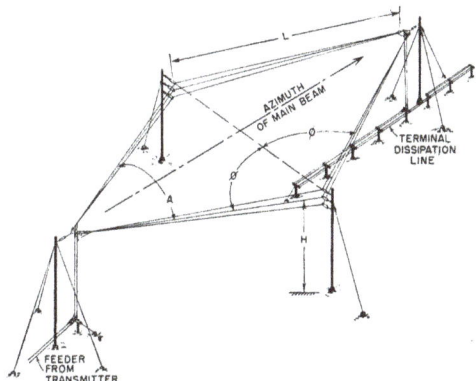

Fig. 3.77. Horizontal rhombic antenna (common three-wire form).

A horizontal three-wire rhombic antenna. This example is terminated with a lossy transmission line instead of a resistor.

AT&T 2 wire rhombic in Dixon, California, in 1937, used for telephone service to Shanghai, China

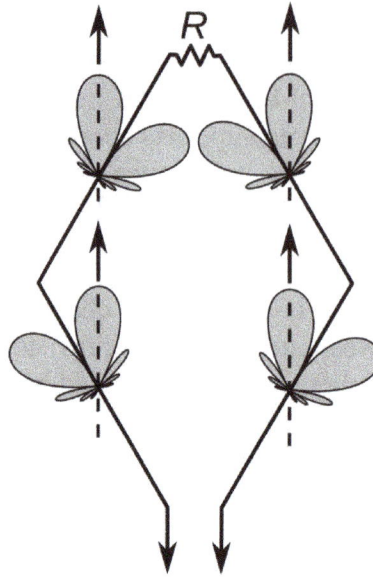

Diagram of radiation patterns (grey) of each segment of the antenna illustrates how it works. By using the correct vertex angle, one of the main lobes of each of the 4 sides point in the same direction, reinforcing each other, increasing the gain.

A rhombic antenna is a broadband directional wire antenna co-invented by Edmond Bruce and Harald Friis, in 1931, mostly commonly used in the high frequency (HF) or shortwave band.

Description

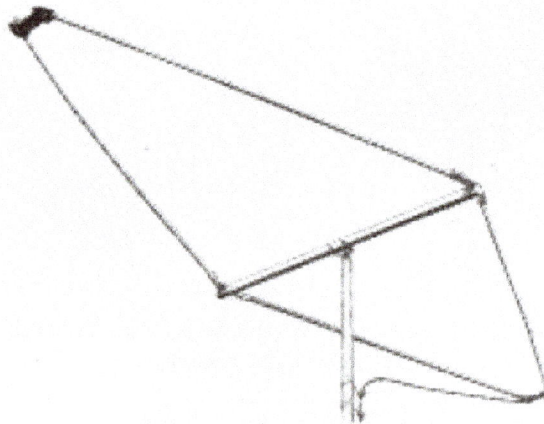

Small rhombic UHF television antenna from 1952. Its broad bandwidth allowed it to cover the 470 to 890 MHz UHF television band.

It consists of one to three parallel wires suspended above the ground in a "rhombic" (diamond) shape, supported by poles or towers at each vertex to which the wires are attached by insulators. Each of the four sides are the same length, typically at least one wavelength (λ) or longer. A horizontal rhombic antenna *(picture, top right)* radiates horizontally polarised waves. Its principal advantages over other types of antenna are its simplicity, high forward gain and wide bandwidth, the ability to operate over a wide range of frequencies.

It is typically fed at one of the two acute (sharper angle) vertices through a balanced transmission line. Less commonly, it can be fed with coaxial cable through a balun transformer. The end of the wires meeting at the opposite vertex is either left open (unconnected), or terminated with a non-inductive resistor. When resistor-terminated, the radiation pattern is unidirectional, with the main lobe off the terminated end, so this end of the antenna is oriented toward the target country or geographical region. When unterminated the rhombic is bidirectional, with two opposite lobes off the two acute ends, but is not perfectly bi-directional. This is because of energy losses caused by radiation, conductor resistance, and coupling to the lossy soil below the antenna.

The rhombic antenna, like other horizontal antennas, can radiate at elevation angles close to the horizon or at higher angles depending on its height above ground relative to the operating frequency and its physical construction. Likewise, its beamwidth can be narrow or broad, depending primarily on its length. The shallow radiation angle makes it useful for skywave ("skip") propagation, the dominant mode at shortwave frequencies, in which radio waves directed at an angle into the sky reflect from layers in the ionosphere and return to Earth beyond the horizon.

A rhombic requires a large area of land — especially if several antennas are installed to serve a variety of geographic regions at different distances or directions or to cover widely different frequencies. The rhombic suffers from efficiency problems due to earth losses below the antenna, significant power-wasting spurious lobes, termination losses, and the inability to maintain constant current along the length of the conductors. Typical radiation efficiency is in the order of 40-50%. The low efficiency significantly reduces gain for a given main lobe beamwidth when compared to other arrays of the same beamwidth.

At the expense of system simplicity, it is possible to improve efficiency by recirculation of power wasted in the termination resistance of unidirectional rhombics. Use of a recirculating termination system can move efficiency into the 70-80% range by combining power that would have been wasted in the termination with the transmitter power. Such systems bring a low-loss balanced line back from the termination end to the feedpoint through a matching and phasing system. Energy that would otherwise be dissipated in the termination resistance is applied in-phase with the excitation.

Prior to WWII, the rhombic was one of the most popular point-to-point high frequency antenna arrays. After WWII the rhombic largely fell out of favor for shortwave broadcast and point-to-point communications work, being replaced by log periodic antennas and curtain arrays. Larger log periodics provide wider frequency coverage with comparable gain to rhombics. Distributed feed curtains or HRS curtain arrays provided a cleaner pattern, ability to steer the pattern in elevation and azimuth, much higher efficiency, and significantly higher gain in less space. However, rhombic antennas are used in cases where the combination of high forward gain (despite the losses described above) and large operating bandwidth cannot be achieved by other means.

The rhombic remains one of the least complex medium-gain options for sustained long distance communications over point-to-point circuits. Rhombics also handle considerable transmitter power, since they have essentially uniform voltage and current distribution. The rhombic's low cost, simplicity, reliability, and ease of construction sometimes outweighs performance advantages offered by other more complex arrays.

Advantages of Rhombic Antennas

Its input impedance & radiation pattern are relatively constant over a 2:1 range of frequencies. Its impedance can be made relatively constant over a frequency range 4:1 or more, with the forward gain increasing at 6 dB per octave.

- Multiple rhombic antennas can be connected in an end-to-end fashion to form MUSA (Multiple Unit Steerable Antenna). MUSA arrays can receive long distance, short wave, horizontally polarized downcoming waves.

- In addition to its use as a simple and effective transmitting antenna (as described above), the rhombic can also be used as an HF receiving antenna with good gain and directivity. For example, BBC Monitoring's Crowsley Park receiving station has three rhombic antennas aligned for reception at azimuths of 37, 57 and 77 degrees.

Helical Antenna

Array of four axial-mode helical antennas used as a satellite tracking-acquisition antenna,Pleumeur-Bodou, France

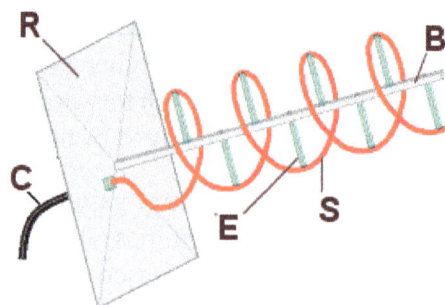

Helical antenna:

(B) Central support,

(C) Coaxial cable feedline,

(E) Insulating supports for the helix,

(R) Reflector ground plane,*(S)* Helical radiating wire

A helical antenna is an antenna consisting of a conducting wire wound in the form of a helix. In most cases, helical antennas are mounted over a ground plane. The feed line is connected between the bottom of the helix and the ground plane. Helical antennas can operate in one of two principal modes — normal mode or axial mode.

In the *normal mode* or *broadside* helix, the dimensions of the helix (the diameter and the pitch) are small compared with the wavelength. The antenna acts similarly to an electrically short dipole or monopole, and the radiation pattern, similar to these antennas is omnidirectional, with maximum radiation at right angles to the helix axis. The radiation is linearly polarised parallel to the helix axis. These are used for compact antennas for portable and mobile two-way radios, and for UHF television broadcasting antennas.

In the *axial mode* or *end-fire* helix, the dimensions of the helix are comparable to a wavelength. The antenna functions as a directional antenna radiating a beam off the ends of the helix, along the antenna's axis. It radiates circularly polarised radio waves. These are used for satellite communication.

Normal-mode Helical

Normal-mode helical UHF TV broadcasting antenna 1954

If the circumference of the helix is significantly less than a wavelength and its *pitch* (axial distance between successive turns) is significantly less than a quarter wavelength, the antenna is called a *normal-mode* helix. The antenna acts similar to a monopole antenna, with an omnidirectional radiation pattern, radiating equal power in all directions perpendicular to the antenna. However, because of the inductance added by the helical shape, the antenna acts like a *inductively loaded* monopole; at its resonant frequency it is shorter than a quarter-wavelength long. Therefore normal-mode helices can be used as electrically short monopoles, an alternative to center- or base-loaded whip antennas, in applications where a full sized quarter-wave monopole would be too big. As with other electrically short antennas, the gain, and thus the communication range, of the helix will be less than that of a full sized antenna. Their compact size makes "helicals" useful as antennas for mobile and portable communications equipment on the HF, VHF, and UHF bands.

A common form of normal-mode helical antenna is the "rubber ducky antenna" used in portable radios.
The loading provided by the helix allows the antenna to be physically shorter than
its electrical length of a quarter-wavelength.

An effect of using a helical conductor rather than a straight one is that the matching impedance is changed from the nominal 50 ohms to between 25 to 35 ohms base impedance. This does not seem to be adverse to operation or matching with a normal 50 ohm transmission line, provided the connecting feed is the electrical equivalent of a 1/2 wavelength at the frequency of operation.

Another example of the type as used in mobile communications is "spaced constant turn" in which two or more different linear windings are wound on a single former and spaced so as to provide an efficient balance between capacitance and inductance for the radiating element at a particular resonant frequency.

Many examples of this type have been used extensively for 27 MHz CB radio with a wide variety of designs originating in the US and Australia in the late 1960s. Multi-frequency versions with plug-in taps have become the mainstay for multi-band Single-sideband modulation (SSB) HF communications.

Most examples were wound with copper wire using a fiberglass rod as a former. This flexible radiator is then covered with heat-shrink tubing which provides a resilient and rugged waterproof covering for the finished mobile antenna.

These popular designs are still in common use as of 2010 and have been universally adapted as standard FM receiving antennas for many factory produced motor vehicles as well as the existing basic style of aftermarket HF and VHF mobile helical. The most common use for broadside helixes is in the "rubber ducky antenna" found on most portable VHF and UHF radios.

Helical Broadcasting Antennas

Specialized normal-mode helical antennas are used for FM radio and television broadcasting on the VHF and UHF bands.

Axial-mode Helical

When the helix circumference is near the wavelength of operation, the antenna operates in *axial mode*. This is a nonresonant traveling wave mode, in which instead of standing waves, the waves of current and voltage travel in one direction, up the helix. Instead of radiating linearly polarized waves normal to the antenna's axis , it radiates a beam of radio waves with circular polarisation along the axis, off the ends of the antenna. The main lobes of the radiation pattern are along the axis of the helix, off both ends. Since in a directional antenna only radiation in one direction is wanted, the other end of the helix is terminated in a flat metal sheet or screen reflector to reflect the waves forward.

End fire helical satellite communications antenna, Scott Air Force base, Illinois, USA. Satellite communication systems often use circularly polarised radio waves, because the satellite antenna may be oriented at any angle in space without affecting the transmission, and axial mode (end fire) helical antennas are often used as the ground antenna.

In radio transmission, circular polarisation is often used where the relative orientation of the transmitting and receiving antennas cannot be easily controlled, such as in animal tracking and spacecraft communications, or where the polarisation of the signal may change, so end-fire helical antennas are frequently used for these applications. Since large helices are difficult to build and unwieldy to steer and aim, the design is commonly employed only at higher frequencies, ranging from VHF up to microwave.

The helix in the antenna can twist in two possible directions: right-handed or left-handed, as defined by the right hand rule. In an axial-mode helical antenna the direction of twist of the helix determines the polarisation of the radio waves: a left-handed helix radiates left-circularly-polarised radio waves, a right-handed helix radiates right-circularly-polarised radio waves. Helical antennas can receive signals with any type of linear polarisation, such as horizontal or vertical polarisation, but when receiving circularly polarised signals the handedness of the receiving antenna must be the same as the transmitting antenna; left-hand polarised antennas suffer a severe loss of gain when receiving right-circularly-polarised signals, and vice versa.

The dimensions of the helix are determined by the wavelength λ of the radio waves used, which depends on the frequency. In order to operate in axial-mode, the circumference should be equal to the wavelength. The pitch angle should be 13 degrees, which is a pitch distance (distance between each turn) of 0.23 times the circumference, which means the spacing between the coils should be approximately one-quarter of the wavelength (λ/4). The number of turns in the helix determines how directional the antenna is: more turns improves the gain in the direction of its axis at both ends (or at 1 end when a ground plate is used), at a cost of gain in the other directions. When C<λ it operates more in normal mode where the gain direction is a donut shape to the sides instead of out the ends.

Terminal impedance in axial mode ranges between 100 and 200 ohms, approximately

$$Z \simeq 140\left(\frac{C}{\lambda}\right)$$

where C is the circumference of the helix, and λ is the wavelength. Impedance matching (when C=λ) to standard 50 or 75 ohm coaxial cable is often done by a quarter wave stripline section acting as an impedance transformer between the helix and the ground plate.

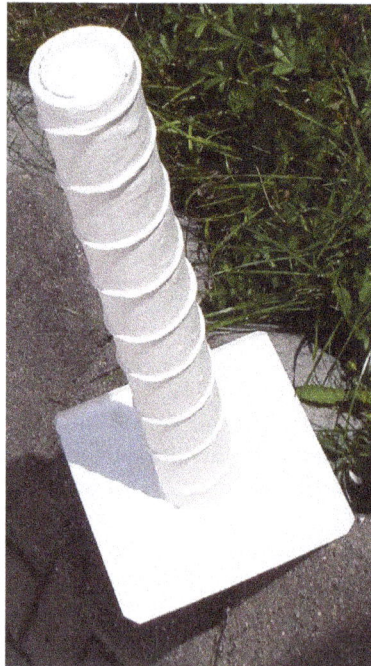

Helical antenna for WLAN communication, working frequency app. 2.4 GHz

The maximum directive gain is approximately:

$$Gain \simeq 15\left(\frac{C}{\lambda}\right)^2\left(\frac{NS}{\lambda}\right)$$

where N is the number of turns and S is the spacing between turns. Most designs use C=λ and S=0.23*C, so the gain is typically G=0.8*N. In decibels, the gain is $G_{dBi} = 10 \cdot \log_{10}(0.8N)$..

The half-power beamwidth is:

$$\text{HPBW} \simeq \frac{52}{\frac{C}{\lambda}\sqrt{\frac{NS}{\lambda}}}\text{degrees}$$

The beamwidth between nulls is:

$$\text{FNBW} \simeq \frac{115\lambda^{3/2}}{C\sqrt{NS}}\text{degrees}$$

Directional Antenna

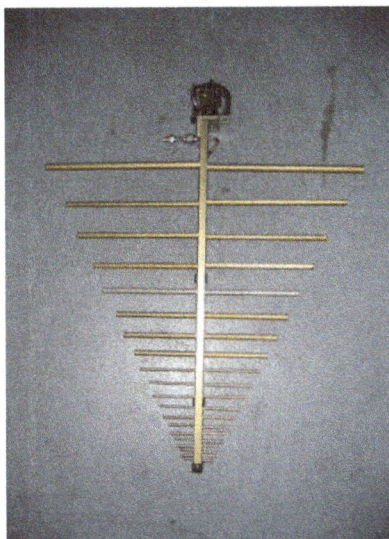

A multi-element, log-periodic dipole array

A directional antenna or beam antenna is an antenna which radiates or receives greater power in specific directions allowing for increased performance and reduced interference from unwanted sources. Directional antennas provide increased performance over dipole antennas – or omnidirectional antennas in general – when greater concentration of radiation in a certain direction is desired.

A high-gain antenna (HGA) is a directional antenna with a focused, narrow radiowave beam width. This narrow beam width allows more precise targeting of the radio signals. Most commonly referred to during space missions, these antennas are also in use all over Earth, most successfully in flat, open areas where no mountains lie to disrupt radiowaves. By contrast, a low-gain antenna (LGA) is an omnidirectional antenna with a broad radiowave beam width, that allows for the signal to propagate reasonably well even in mountainous regions and is thus more reliable regardless of terrain. Low gain antennas are often used in spacecraft as a backup to the *high-gain antenna*, which transmits a much narrower beam and is therefore susceptible to loss of signal.

A 70-meter Cassegrain radio antenna at GDSCC, California

All practical antennas are at least somewhat directional, although usually only the direction in the plane parallel to the earth is considered, and practical antennas can easily be omnidirectional in one plane. The most common types are the Yagi antenna, the log-periodic antenna, and the corner reflector antenna, which are frequently combined and commercially sold as residential TV antennas. Cellular repeaters often make use of external directional antennas to give a far greater signal than can be obtained on a standard cell phone. Satellite Television receivers usually use parabolic antennas. For long and medium wavelength frequencies, tower arrays are used in most cases as directional antennas.

Principle of Operation

When transmitting, a *high-gain antenna* allows more of the transmitted power to be sent in the direction of the receiver, increasing the received signal strength. When receiving, a high gain antenna captures more of the signal, again increasing signal strength. Due to reciprocity, these two effects are equal - an antenna that makes a transmitted signal 100 times stronger (compared to an isotropic radiator), will also capture 100 times as much energy as the isotropic antenna when used as a receiving antenna. As a consequence of their directivity, directional antennas also send less (and receive less) signal from directions other than the main beam. This property may be used to reduce interference.

There are many ways to make a high-gain antenna - the most common are parabolic antennas, helical antennas, yagi antennas, and phased arrays of smaller antennas of any kind. Horn antennas can also be constructed with high gain, but are less commonly seen. Still other configurations are possible - the Arecibo Observatory uses a combination of a *line feed* with an enormous spherical reflector (as opposed to a more usual parabolic reflector), to achieve extremely high gains at specific frequencies.

Antenna Gain

Antenna gain is often quoted with respect to a hypothetical antenna that radiates equally in all directions, an isotropic radiator. This gain, when measured in decibels, is called dBi. Conservation of energy dictates that high gain antennas must have narrow beams. For example, if a high gain antenna makes a 1 watt transmitter look like a 100 watt transmitter, then the beam can cover at most 1/100 of the sky (otherwise the total amount of energy radiated in all directions would sum to more than the transmitter power, which is not possible). In turn this implies that high-gain antennas must be physically large, since according to the diffraction limit, the narrower the beam desired, the larger the antenna must be (measured in wavelengths).

Antenna gain can also be measured in dBd, which is gain in Decibels compared to the maximum intensity direction of a half wave dipole. In the case of Yagi type aerials this more or less equates to the gain one would expect from the aerial under test minus all its directors and reflector. It is important not to confuse dBi and dBd; the two differ by 2.15 dB, with the dBi figure being higher, since a dipole has 2.15 db of gain with respect to an isotropic antenna.

Gain is also dependent on the number of elements and the tuning of those elements. Antennas can be tuned to be resonant over a wider spread of frequencies but, all other things being equal, this will mean the gain of the aerial is lower than one tuned for a single frequency or a group of frequencies. For example, in the case of wideband TV antennas the fall off in gain is particularly large at the bottom of the TV transmitting band. In the UK this bottom third of the TV band is known as group A, see gain graph comparing grouped aerials to a wideband aerial of the same size/model.

Other factors may also affect gain such as aperture (the area the antenna collects signal from, almost entirely related to the size of the antenna but for small antennas can be increased by adding a ferrite rod), and efficiency (again, affected by size, but also resistivity of the materials used and impedance matching). These factors are easy to improve without adjusting other features of the antennas or coincidentally improved by the same factors that increase directivity, and so are typically not emphasized.

Applications

High gain antennas are typically the largest component of deep space probes, and the highest gain radio antennas are physically enormous structures, such as the Arecibo Observatory. The Deep Space Network uses 35 meter dishes at about 1 cm wavelengths. This combination gives the antenna gain of about 100,000,000 (or 80 dB, as normally measured), making the transmitter appear about 100 million times stronger, and a receiver about 100 million times more sensitive, *provided the target is within the beam*. This beam can cover at most 1/100 millionth of the sky, so very accurate pointing is required.

Television Antenna

A television antenna, or TV aerial, is an antenna specifically designed for the reception of over-the-air broadcast television signals, which are transmitted at frequencies from about 41 to 250 MHz in the VHF band, and 470 to 960 MHz in the UHF band in different countries. Television antennas are manufactured in two different types: "indoor" antennas, to be located on top of or next to the television set, and "outdoor" antennas, mounted on a mast on top of the owner's house. The most common types of antennas used are the dipole ("rabbit ears") and loop antennas, and for outdoor antennas the yagi and log periodic, and UHF multi-bay "flat" (Reflective array antenna) antennae.

A Winegard 68 element VHF/UHF aerial antenna. This common multi-band antenna type uses a UHF yagi at the front and a VHF log-periodic at the back coupled together.

Description

To cover this range, antennas generally consist of multiple conductors of different lengths which correspond to the wavelength range the antenna is intended to receive. The length of the elements of a TV antenna are usually half the wavelength of the signal they are intended to receive. The wavelength of a signal equals the speed of light (c) divided by the frequency. The design of a television broadcast receiving antenna is the same for the older analog transmissions and the digital television (DTV) transmissions which are replacing them. Sellers often claim to supply a special "digital" or "high-definition television" (HDTV) antenna advised as a replacement for an existing analog television antenna; at best this is misinformation to generate sales of unneeded equipment, at worst it may leave the viewer with a UHF-only antenna in a local market (particularly in North America) where some digital stations remain on their original high VHF frequencies.

Indoor

Simple half-wave dipole antenna for VHF or UHF loop antennas that are made to be placed indoors are often used for television (and VHF radio); these are often called "rabbit ears" or "bunny aerials". because of their appearance. The length of the telescopic "ears" can be adjusted by the user, and should be about one half of the wavelength of the signal for the desired channel. These are not as efficient as an aerial rooftop antenna since they are less directional and not always adjusted to the proper length for the desired channel. Dipole antennas are bi-directional, that is, they receive evenly forward and backwards, and also cover a broader band than antennas with more elements. This makes them less efficient than antennas designed to maximize the signal from a narrower angle in one direction. Coupled with the poor placing, indoors and closer to the ground, they are much worse than multi-element rooftop antennas at receiving signals which are not very strong, although often adequate for nearby transmitters, in which case they may be adequate and cheap. These simple antennas are called set-top antennas because they were often placed on top of the television set or receiver.

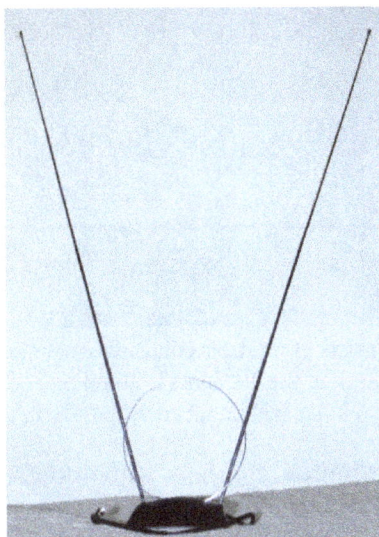

Very common "rabbit ears" set-top antenna.

The actual length of the ears is optimally about 91% of half the wavelength of the desired channel in free space. Quarter-wave television antennas are also used. These use a single element, and use the earth as a ground plane; therefore, no ground is required in the feed line.

Soon after television broadcasting switched from analog to digital broadcasting, indoor antennas have evolved beyond the traditional "rabbit ears." RCA is one manufacturer which has commercially sold a flat antenna. Flat antennas are very lightweight, very thin, and square-shaped like a thin notebook. They connect to televisions, or to digital converter boxes, with a single coaxial cable, and may be sold with an optional signal amplifier. The amplifier must be plugged into a power source, but the flat antenna does not require a power source. The flat antenna may need some moving around to achieve an optimum reception, but it eliminates a lot of manual manipulation which is inherent in use of the "rabbit ears".

Outdoor

In a combination (combo) VHF/UHF antenna the longer elements (for picking up VHF frequencies) are at the "back" of the antenna, relative to the device's directionality, and the much shorter UHF elements are in the "front", and the antenna works best when "pointing" to the source of the signal to be received. The smallest elements in this design, located in the "front", are UHF and use Yagi antenna principles. The longest elements, located in the "back" of the antenna use VHF Log-periodic principles. Combining these two types of antenna creates the combination VHF/UHF antenna commonly used.

An antenna can have a smaller or larger number of directors; the more directors it has (requiring a longer boom), and the more accurate their beamwidth the higher its gain will be. For the commonly used Yagi antenna this is not a linear relationship. Antenna gain is the ratio of the signal received from the preferred direction to the signal from an ideal omnidirectional antenna. Gain is inversely proportional to the antenna's acceptance angle.

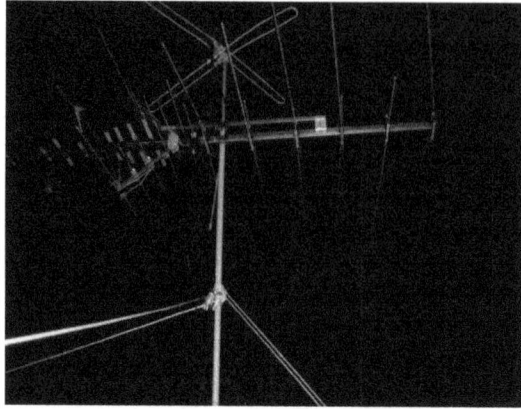

An outdoor TV antenna generally consists of multiple conductive elements that are arranged such that it is a directional antenna. The length of the elements is about one half of the signal wavelength. Therefore, the length of each element corresponds to a certain frequency.

Two or more directional rooftop antennas can be set up and connected to one receiver. Antennas designed for rooftop use are sometimes located in attics.

Sometimes television transmitters are organised such that all receivers in a given location need receive transmissions in only a relatively narrow band of the full UHF television spectrum and from the same direction, so that a single antenna provides reception from all stations.

A UHF television antenna

An antenna pole setup in a chimney, reaching 35 feet (10.7 meters) off the ground

Installation

A short antenna pole next to a house.

Multiple Yagi TV aerials

Antennas are commonly placed on rooftops, and sometimes in attics. Placing an antenna indoors significantly attenuates the level of the available signal. Directional antennas must be pointed at the transmitter they are receiving; in most cases great accuracy is not needed. In a given region it is sometimes arranged that all television transmitters are located in roughly the same direction and use frequencies spaced closely enough that a single antenna suffices for all. A single transmitter location may transmit signals for several channels. CABD (communal antenna broadcast distribution) is a system installed inside a building to receive free-to-air TV/FM signals transmitted via radio frequencies and distribute them to the audience.

Analog television signals are susceptible to ghosting in the image, multiple closely spaced images giving the impression of blurred and repeated images of edges in the picture. This is due to the signal being reflected from nearby objects (buildings, tree, mountains); several copies of the signal, of different strengths and subject to different delays, are picked up. This is different for different transmissions. Careful positioning of the antenna can produce a compromise position which minimizes the ghosts on different channels. Ghosting is also possible if multiple antennas connected to the same receiver pick up the same station, especially if the lengths of the cables connecting them to the splitter/merger are different lengths or the antennas are too close together. Analog television is being replaced by digital, which is not subject to ghosting.

Rooftop and Other Outdoor Antennas

Aerials are attached to roofs in various ways, usually on a pole to elevate it above the roof. This is generally sufficient in most areas. In some places, however, such as a deep valley or near taller structures, the antenna may need to be placed significantly higher, using a guide mast or mast. The wire connecting the antenna to indoors is referred to as the *downlead* or *drop*, and the longer the downlead is, the greater the signal degradation in the wire. If you replace the old cable (wire) with RG6 Quad Shield cabling it minimizes signal loss and lasts longer in severe weather conditions and everyday wear and tear.

The higher the antenna is placed, the better it will perform. An antenna of higher gain will be able to receive weaker signals from its preferred direction. Intervening buildings, topographical features (mountains), and dense forest will weaken the signal; in many cases the signal will be reflected such that a usable signal is still available. There are physical dangers inherent to high or complex antennas, such as the structure falling or being destroyed by weather. There are also varying local ordinances which restrict and limit such things as the height of a structure without obtaining permits. For example, in the United States, the Telecommunications Act of 1996 allows any homeowner to install "An antenna that is designed to receive local television broadcast signals", but that "masts higher than 12 feet above the roof-line may be subject to local permitting requirements."

Indoor Antennas

As discussed previously, antennas may be placed indoors where signals are strong enough to overcome antenna shortcomings. The antenna is simply plugged into the television receiver and placed conveniently, often on the top of the receiver ("set-top"). Sometimes the position needs to be experimented with to get the best picture. Indoor antennas can also benefit from RF amplification, commonly called a TV booster. Indoor antennas will never be an option in weak signal areas.

Attic installationSometimes it is desired not to put an antenna on the roof; in these cases, antennas designed for outdoor use are often mounted in the attic or loft, although antennas designed for attic use are also available. Putting an antenna indoors significantly decreases its performance due to lower elevation above ground level and intervening walls; however, in strong signal areas reception may be satisfactory. One layer of asphalt shingles, roof felt, and a plywood roof deck is considered to attenuate the signal to about half.

Multiple Antennas, Rotators

It is sometimes desired to receive signals from transmitters which are not in the same direction. This can be achieved, for one station at a time, by using a rotator operated by an electric motor to turn the antenna as desired. Alternatively, two or more antennas, each pointing at a desired transmitter and coupled by appropriate circuitry, can be used. To prevent the antennas from interfering with each other, the vertical spacing between the booms must be at least half the wavelength of the lowest frequency to be received (Distance=$\lambda/2$). The wavelength of 54 MHz (Channel 2) is 5.5 meters (λ x f = c) so the antennas must be a minimum of 2.25 metres, or about 89 inches apart. It is also important that the cables connecting the antennas to the signal splitter/merger be exactly the same length, to prevent phasing issues, which cause ghosting with analog reception. That is, the antennas might both pick up the same station; the signal from the one with the shorter cable will reach the receiver slightly sooner, supplying the receiver with two pictures slightly offset. There may be phasing issues even with the same length of down-lead cable. Bandpass filters or "signal traps" may help to reduce this problem.

Two aerials set up on a roof. Spaced horizontally and vertically

For side-by-side placement of multiple antennas, as is common in a space of limited height such as an attic, they should be separated by at least one full wavelength of the lowest frequency to be received at their closest point.

Often when multiple antennas are used, one is for a range of co-located stations and the other is for a single transmitter in a different direction.

Faraday Shielding Your Antenna

You can enhance any antenna rig by creating a Faraday shield on all sides of the antenna but the one towards the source. If you are in range of the transmitter, and your rig is underperforming, it may be due to competition with similar signals. When you put a metal screen between this interference and your rig, it can enhance performance passively, requiring no power source. The screening need not be thick to work, theoretically a molecule thick would do. As long as the interference is in a different line of sight than the transmitter, this works fine.

Safety

- TV antennas are good conductors of electricity and attract lightning, acting as a lightning rod. The use of a lightning arrestor is usual to protect against this. A large grounding rod connected to both the antenna and the mast or pole is required.

- Properly installed masts, especially tall ones, are guyed with galvanized cable; no insulators are needed. They are designed to withstand worst-case weather conditions in the area, and positioned so that they do not interfere with power lines if they fall.

- There is inherent danger in being on the rooftop of a house, required for installing or adjusting a television antenna. British entertainer Rod Hull died after falling from his roof where he had been trying to improve reception for a football match.

Smart Antenna

Smart antennas (also known as adaptive array antennas, multiple antennas and, recently, MIMO) are antenna arrays with smart signal processing algorithms used to identify spatial signal signature such as the direction of arrival (DOA) of the signal, and use it to calculate beamforming vectors, to track and locate the antenna beam on the mobile/target. Smart antennas should not be confused with reconfigurable antennas, which have similar capabilities but are single element antennas and not antenna arrays.

Smart antenna techniques are used notably in acoustic signal processing, track and scan radar, radio astronomy and radio telescopes, and mostly in cellular systems like W-CDMA, UMTS, and LTE.

Smart antennas have two main functions: DOA estimation and Beamforming.

Direction of Arrival (DOA) Estimation

The smart antenna system estimates the direction of arrival of the signal, using techniques such as MUSIC (MUltiple SIgnal Classification), estimation of signal parameters via rotational invariance techniques (ESPRIT) algorithms, Matrix Pencil method or one of their derivatives. They involve finding a spatial spectrum of the antenna/sensor array, and calculating the DOA from the peaks of this spectrum. These calculations are computationally intensive.

Matrix Pencil is very efficient in case of real time systems, and under the correlated sources.

Beamforming

Beamforming is the method used to create the radiation pattern of the antenna array by adding constructively the phases of the signals in the direction of the targets/mobiles desired, and nulling the pattern of the targets/mobiles that are undesired/interfering targets. This can be done with a simple Finite Impulse Response (FIR) tapped delay line filter. The weights of the FIR filter may also be changed adaptively, and used to provide optimal beamforming, in the sense that it reduces the Minimum Mean Square Error between the desired and actual beampattern formed. Typical algorithms are the steepest descent, and Least Mean Squares algorithms.

Types of Smart Antennas

Two of the main types of smart antennas include switched beam smart antennas and adaptive array smart antennas. Switched beam systems have several available fixed beam patterns. A decision is made as to which beam to access, at any given point in time, based upon the requirements of the system. Adaptive arrays allow the antenna to steer the beam to any direction of interest while simultaneously nulling interfering signals. Beamdirection can be estimated using the so-called direction-of-arrival (DOA) estimation methods.

In 2008, the United States NTIA began a major effort to assist consumers in the purchase of digital television converter boxes. Through this effort, many people have been exposed to the concept of smart antennas for the first time. In the context of consumer electronics, a "smart antenna" is one that conforms to the EIA/CEA-909 Standard Interface.

Limited Choice of EIA/CEA-909A Smart Antennas in the Marketplace

Prior to the final transition to ATSC digital television in the United States on June 11, 2009, two smart antenna models were brought to market:

- RCA ANT2000 – no longer available from retailers

- DTA-5000 – manufactured by Funai Electric, marketed under the "DX Antenna" brand name, sometimes associated with the Sylvania brand name; no longer available from retailers

And two models are causing consumer confusion:

- Although the Apex SM550 is capable of connecting to a CEA-909 port for the purpose of drawing electrical power, it is not a true smart antenna.

- The unfortunately-named Channel Master 3000A and CM3000HD SMARTenna series are otherwise-conventional amplified omnidirectional antennas, not steerable smart antennas.

Extension of Smart Antennas

Smart antenna systems are also a defining characteristic of MIMO systems, such as the IEEE 802.11n standard. Conventionally, a smart antenna is a unit of a wireless communication system and performs spatial signal processing with multiple antennas. Multiple antennas can be used at either the transmitter or receiver. Recently, the technology has been extended to use the multiple antennas at both the transmitter and receiver; such a system is called a multiple-input multiple-output (MIMO) system. As extended Smart Antenna technology, MIMO supports spatial information processing, in the sense that conventional research on Smart Antennas has focused on how to provide a beamforming advantage by the use of spatial signal processing in wireless channels. Spatial information processing includes spatial information coding such as Spatial multiplexing and Diversity Coding, as well as beamforming.

Reflective Array Antenna

In telecommunications and radar, a reflective array antenna is a class of directive antennas in which multiple driven elements are mounted in front of a flat surface designed to reflect the radio waves in a desired direction. They are often used in the VHF and UHF frequency bands. VHF examples are generally large and resemble a highway billboard, so they are sometimes called billboard antennas or broadside antennas. Other names are bedspread array and bowtie array depending on the type of elements making up the antenna. The curtain array is a larger version used by shortwave radio broadcasting stations.

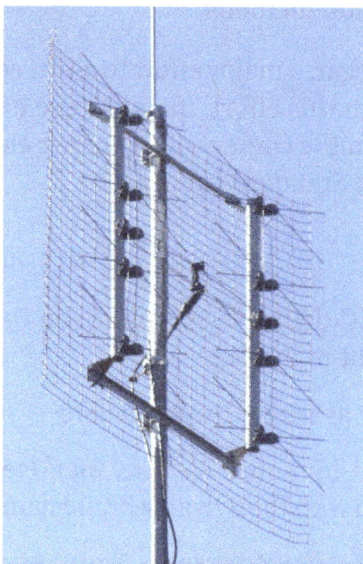

This reflective array television antenna consists of eight "bowtie" dipole driven elements mounted in frontof a wire screen reflector.The X-shaped dipoles give it a wide bandwidth to cover both the VHF (174-216 MHz)and UHF (470-700 MHz) bands. It has a gain of 5 dB VHF and 12 dB UHF and an 18 dB front-to-back ratio.

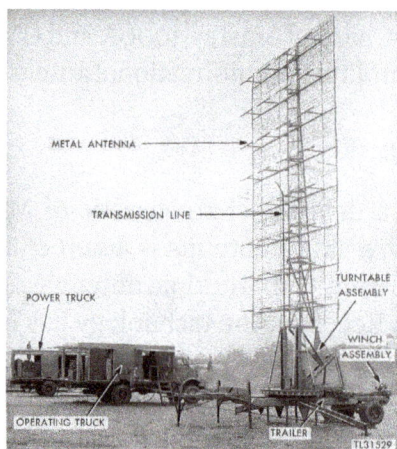

Reflective array 'billboard' antenna of the AN-270 radar, an early US Army radar system. It consists of 32 horizontal half wave dipoles mounted in front of a 55 ft. high screen reflector. With an operat w beamwidth to locate enemy aircraft.

Reflective array antennas usually have a number of identical driven elements, fed in phase, in front of a flat, electrically large reflecting surface to produce a unidirectional beam, increasing antenna gain and reducing radiation in unwanted directions. The individual elements are most commonly half wave dipoles, although they sometimes contain parasitic elements as well as driven elements. The reflector may be a metal sheet or more commonly a wire screen. A metal screen reflects radio waves as well as a solid metal sheet as long as the holes in the screen are smaller than about one-tenth of a wavelength, so screens are often used to reduce weight and wind loads on the antenna. They usually consist of a grill of parallel wires or rods, oriented parallel to the axis of the dipole elements.

The driven elements are fed by a network of transmission lines, which divide the power from the RF source equally between the elements. This often has the circuit geometry of a tree structure.

Basic Concepts

Radio Signals

When a radio signal passes a conductor, it induces an electrical current in it. Since the radio signal fills space, and the conductor has a finite size, the induced currents add up or cancel out as they move along the conductor. A basic goal of antenna design is to make the currents add up to a maximum at the point where the energy is tapped off. To do this, the antenna elements are sized in relation to the wavelength of the radio signal, with the aim of setting up standing waves of current that are maximized at the feed point.

This means that an antenna designed to receive a particular wavelength has a natural size. To improve reception, one cannot simply make the antenna larger; this will improve the amount of signal intercepted by the antenna, which is largely a function of area, but will lower the efficiency of the reception (at a given wavelength). Thus, in order to improve reception, antenna designers often use multiple elements, combining them together so their signals add up. These are known as *antenna arrays*.

Array Phasing

In order for the signals to add together, they need to arrive in-phase. Consider two dipole antennas placed in a line end-to-end, or *collinear*. If the resulting array is pointed directly at the source signal, both dipoles will see the same instantaneous signal, and thus their reception will be in-phase. However, if one were to rotate the antenna so it was at an angle to the signal, the extra path from the signal to the more distant dipole means it receives the signal slightly out of phase. When the two signals are then added up, they no longer strictly reinforce each other, and the output drops. This makes the array more sensitive horizontally, while stacking the dipoles in parallel narrows the pattern vertically. This allows the designer to tailor the reception pattern, and thus the gain, by moving the elements about.

If the antenna is properly aligned with the signal, at any given instant in time, all of the elements in an array will receive the same signal and be in-phase. However, the output from each element has to be gathered up at a single feed point, and as the signals travel across the antenna to that point, their phase is changing. In a two-element array this is not a problem because the feed point can be placed between them; any phase shift taking place in the transmission lines is equal for both elements. However, if one extends this to a four-element array, this approach no longer works, as the signal from the outer pair has to travel further and will thus be at a different phase than the inner pair when it reaches the center. To ensure that they all arrive with the same phase, it is common to see additional transmission wire inserted in the signal path, or for the transmission line to be crossed over to reverse the phase if the difference is greater than 1/2 a wavelength.

Reflectors

The gain can be further improved through the addition of a *reflector*. Generally any conductor in a flat sheet will act in a mirror-like fashion for radio signals, but this also holds true for non-continuous surfaces as long as the gaps between the conductors are less than about 1⁄10 of the target wavelength. This means that wire mesh or even parallel wires or metal bars can be used, which is especially useful both for reducing the total amount of material as well as reducing wind loads.

Due to the change in signal propagation direction on reflection, the signal undergoes a reversal of phase. In order for the reflector to add to the output signal, it has to reach the elements in-phase. Generally this would require the reflector to be placed at 1⁄2 of a wavelength behind the elements, and this can be seen in many common reflector arrays like television antennas. However, there are a number of factors that can change this distance, and actual reflector positioning varies.

Reflectors also have the advantage of reducing the signal received from the back of the antenna. Signals received from the rear and re-broadcast from the reflector have not undergone a change of phase, and do not add to the signal from the front. This greatly improves the front-to-back ratio of the antenna, making it more directional. This can be useful when a more directional signal is desired, or unwanted signals are present. There are cases when this is not desirable, and although reflectors are commonly seen in array antennas, they are not universal. For instance, while UHF television antennas often use an array of bowtie antennas with a reflector, a bowtie array without a reflector is a relatively common design in the microwave region.

Gain Limits

As more elements are added to an array, the beamwidth of the antenna's main lobe decreases, leading to an increase in gain. In theory there is no limit to this process. However, as the number of elements increases, the complexity of the required feed network that keeps the signals in-phase increases. Ultimately, the rising inherent losses in the feed network become greater than the additional gain achieved with more elements, limiting the maximum gain that can be achieved.

The gain of practical array antennas is limited to about 25 - 30 dB. Common 4-bay television antennas have gains around 10 to 12 dB, and 8-bay designs might increase this to 12 to 16 dB. The 32-element SCR-270 had a gain around 19.8 dB. Some very large reflective arrays have been constructed, notably the Soviet Duga radars which are hundreds of meters across and contain hundreds of elements. *Active* array antennas, in which groups of elements are driven by separate RF amplifiers, can have much higher gain, but are prohibitively expensive.

Since the 1980s, versions for use at microwave frequencies have been made with patch antenna elements mounted in front of a metal surface.

Radiation Pattern and Beam Steering

When driven in phase, the radiation pattern of the reflective array is a single main lobe perpendicular to the plane of the antenna, plus several sidelobes at equal angles to either side. The more elements used, the narrower the main lobe and the less power is radiated in the sidelobes.

The main lobe of the antenna can be steered electronically within a limited angle by phase shifting the drive signals applied to the individual elements. Each antenna element is fed through a phase shifter which can be controlled digitally, delaying each signal by a successive amount. This causes the wavefronts created by the superposition of the individual elements to be at an angle to the plane of the antenna. Antennas that use this technique are called phased arrays and are being intensively developed, particularly for use in radar systems.

Another option for steering the beam is mounting the entire array structure on a rotating bearing and rotating it mechanically.

Metamaterial Antenna

Metamaterial antennas are a class of antennas which use metamaterials to increase performance of miniaturized (electrically small) antenna systems. Their purpose, as with any electromagnetic antenna, is to launch energy into free space. However, this class of antenna incorporates metamaterials, which are materials engineered with novel, often microscopic, structures to produce unusual physical properties. Antenna designs incorporating metamaterials can step-up the antenna's radiated power.

Conventional antennas that are very small compared to the wavelength reflect most of the signal back to the source. A metamaterial antenna behaves as if it were much larger than its actual size, because its novel structure stores and re-radiates energy. Established lithography techniques can be used to print metamaterial elements on a PC board.

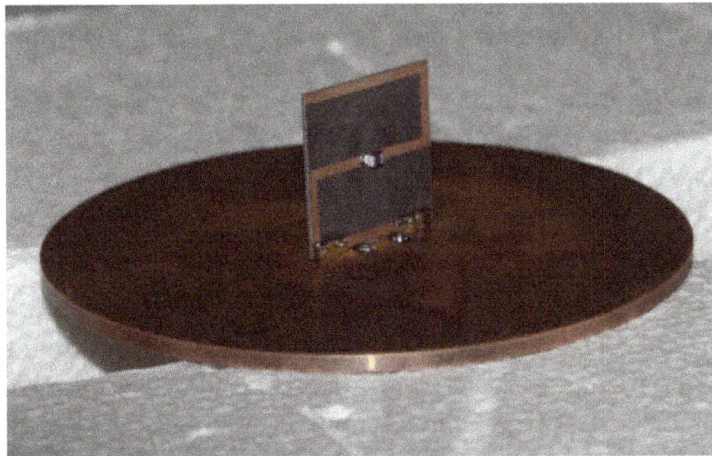

This Z antenna tested at the National Institute of Standards and Technology is smaller than a standard antenna with comparable properties. Its high efficiency is derived from the "Z element" inside the square that acts as a metamaterial, greatly boosting the radiated signal. The square is 30 millimeters on a side.

These novel antennas aid applications such as portable interaction with satellites, wide angle beam steering, emergency communications devices, micro-sensors and portable ground-penetrating radars to search for geophysical features.

Some applications for metamaterial antennas are wireless communication, space communications, GPS, satellites, space vehicle navigation and airplanes.

Antennas Designs

Antenna designs incorporating metamaterials can step-up the radiated power of an antenna. The newest metamaterial antennas radiate as much as 95 percent of an input radio signal. Standard antennas need to be at least half the size of the signal wavelength to operate efficiently. At 300 MHz, for instance, an antenna would need to be half a meter long. In contrast, experimental metamaterial antennas are as small as one-fiftieth of a wavelength, and could have further decreases in size.

Metamaterials are a basis for further miniaturization of microwave antennas, with efficient power and acceptable bandwidth. Antennas employing metamaterials offer the possibility of overcoming restrictive efficiency-bandwidth limitations for conventionally constructed, miniature antennas.

Metamaterials permit smaller antenna elements that cover a wider frequency range, thus making better use of available space for space-constrained cases. In these instances, miniature antennas with high gain are significantly relevant because the radiating elements are combined into large antenna arrays. Furthermore, metamaterials' negative refractive index focuses electromagnetic radiation by a flat lens versus being dispersed.

The DNG Shell

The earliest research in metamaterial antennas was an analytical study of a miniature dipole antenna surrounded with a metamaterial. This material is known variously as a negative index metamaterial (NIM) or double negative metamaterial (DNG) among other names.

This configuration analytically and numerically appears to produce an order of magnitude increase in power. At the same time, the reactance appears to offer a corresponding decrease. Furthermore, the DNG shell becomes a natural impedance matching network for this system.

Ground Plane Applications

Metamaterials employed in the ground planes surrounding antennas offer improved isolation between radio frequency, or microwave channels of (multiple-input multiple-output) (MIMO) antenna arrays. Metamaterial, high-impedance groundplanes can also improve radiation efficiency and axial ratio performance of low-profile antennas located close to the ground plane surface. Metamaterials have also been used to increase beam scanning range by using both the forward and backward waves in leaky wave antennas. Various metamaterial antenna systems can be employed to support surveillance sensors, communication links, navigation systems and command and control systems.

Novel Configurations

Besides antenna miniaturization, the novel configurations have potential applications ranging from radio frequency devices to optical devices. Other combinations, for other devices in metamaterial antenna subsystems are being researched. Either double negative metamaterial slabs are used exclusively or combinations of double positive (DPS) with DNG slabs, or epsilon-negative (ENG) slabs with mu-negative (MNG) slabs are employed in the subsystems. Antenna subsystems that are currently being researched include cavity resonators, waveguides, scatters and antennas (radiators). Metamaterial antennas were commercially available by 2009.

History

Pendry *et al.* were able to show that a three-dimensional array of intersecting, thin wires could be used to create negative values of permittivity (or "ε"), and that a periodic array of copper split ring resonators could produce an effective negative magnetic permeability (or "μ").

In May 2000, a group of researchers, Smith *et al.* were the first to successfully combine the split-ring resonator (SRR), with thin wire conducting posts and produce a left-handed material that had negative values of ε, μ and refractive index for frequencies in the gigahertz or microwave range.

In 2002, a different class of negative refractive index (NRI) metamaterials was introduced that employs periodic reactive loading of a 2-D transmission line as the host medium. This configuration used positive index (DPS) material with negative index material (DNG). It employed a small, planar, negative-refractive-lens interfaced with a positive index, parallel-plate waveguide. This was experimentally verified soon after.

Although some SRR inefficiencies were identified, they continued to be employed as of 2009 for research. SRRs have been involved in wide ranging metamaterial research, including research on metamaterial antennas.

A more recent view is that by using SRRs as building blocks, the electromagnetic response and associated flexibility is practical and desirable.

Phase Compensation Due to Negative Refraction

DNG can provide phase compensation due to their negative index of refraction. This is accomplished by combining a slab of conventional lossless DPS material with a slab of lossless DNG metamaterial.

DPS has a conventional positive index of refraction, while the DNG has a negative refractive index. Both slabs are impedance-matched to the outside region (e.g., free space). The desired monochromatic plane wave is radiated on this configuration. As this wave propagates through the first slab of material a phase difference emerges between the exit and entrance faces. As the wave propagates through the second slab the phase difference is significantly decreased and even compensated for. Therefore, as the wave exits the second slab the total phase difference is equal to zero.

With this system a phase-compensated, waveguiding system could be produced. By stacking slabs of this configuration, the phase compensation (beam translation effects) would occur throughout the entire system. Furthermore, by changing the index of any of the DPS-DNG pairs, the speed at which the beam enters the front face, and exits the back face of the entire stack-system changes. In this manner, a volumetric, low loss, time delay transmission line could be realized for a given system.

Furthermore, this phase compensation can lead to a set of applications, which are miniaturized, subwavelength, cavity resonators, and waveguides with applications below diffraction limits.

Transmission Line Dispersion Compensation

Because of DNG's dispersive nature as a transmission medium, it could be useful as a dispersion compensation device for time-domain applications. The dispersion produces a variance of the group speed of the signals' wave components, as they propagate in the DNG medium. Hence, stacked DNG metamaterials could be useful for modifying signal propagation along a microstrip transmission line. At the same time, dispersion leads to distortion. However, if the dispersion could be compensated for along the microstrip line, RF or microwave signals propagating along them would significantly decrease distortion. Therefore, components for attenuating distortion become less critical, and could lead to simplification of many systems. Metamaterials can eliminate dispersion along the microstrip by correcting for the frequency dependence of the effective permittivity.

The strategy is to design a length of metamaterial-loaded transmission line that can be introduced with the original length of microstrip line to make the paired system dispersionless creating a dispersion-compensating segment of transmission line. This could be accomplished by introducing a metamaterial with a specific localized permittivity and a specific localized magnetic permeability, which then affects the relative permittivity and permeability of the overall microstrip line. It is introduced so that the wave impedance in the metamaterial remains unhanged. The index of refraction in the medium compensates for the dispersion effects associated with the microstrip geometry itself; making the effective refractive index of the pair that of free space.

Part of the design strategy is that the effective permittivity and permeability of such a metamaterial should be negative – requiring a DNG material.

Innovation

Combining left-handed segments with a conventional (right-handed) transmission line results in advantages over conventional designs. Left-handed transmission lines are essentially a high-pass filter with phase advance. Conversely, right-handed transmission lines are a low-pass filter with phase lag. This configuration is designated composite right/left-handed (CRLH) metamaterial.

The conventional Leaky Wave antenna has had limited commercial success because it lacks complete backfire-to-endfire frequency scanning capability. The CRLH allowed complete backfire-to-endfire frequency scanning, including broadside.

Microwave Lens

The metamaterial lens, found in metamaterial antenna systems, is used as an efficient coupler to external radiation, focusing radiation along or from a microstrip transmission line into transmitting and receiving components. Hence, it can be used as an input device. In addition, it can enhance the amplitude of evanescent waves, as well as correct the phase of propagating waves.

Directing Radiation

In this instance an SRR uses layers of a metallic mesh of thin wires – with wires in the three directions of space and slices of foam. This material's permittivity above the plasma frequency can be positive and less than one. This means that the refractive index is just above zero. The relevant parameter is often the contrast between the permittivities rather than the overall permittivity value at desired frequencies. This occurs because the equivalent (effective) permittivity has a behavior governed by a plasma frequency in the microwave domain. This low optical index material then is a good candidate for extremely convergent microlenses. Methods that have been developed theoretically using dielectric photonic crystals applied in the microwave domain to realize a directive emitter using metallic grids.

In this instance, arrayed wires in a cubic, crystal lattice structure can be analyzed as an array of aerials (antenna array). As a lattice structure it has a lattice constant. The lattice constant or lattice parameter refers to the constant distance between unit cells in a crystal lattice.

The earlier discovery of plasmons created the view that metal at plasmon frequency f_p is a composite material. The effect of plasmons on any metal sample is to create properties in the metal such that it can behave as a dielectric, independent of the wave vector of the EM excitation (radiation) field. Furthermore, a minute-fractionally small amount of plasmon energy is absorbed into the system denoted as γ. For aluminium f_p = 15 eV, and γ = 0.1 eV. Perhaps the most important result of the interaction of metal and the plasma frequency is that permittivity is negative below the plasma frequency, all the way to the minute value of γ.

These facts ultimately result in the arrayed wire structure as being effectively a homogeneous medium.

This metamaterial allows for control of the direction of emission of an electromagnetic radiation source located inside the material in order to collect all the energy in a small angular domain around the normal. By using a slab of a metamaterial, diverging electromagnetic waves are focused into a narrow cone. Dimensions are small in comparison to the wavelength and thus the slab behaves as a homogeneous material with a low plasma frequency.

Transmission Line Models

conventional Transmission Lines

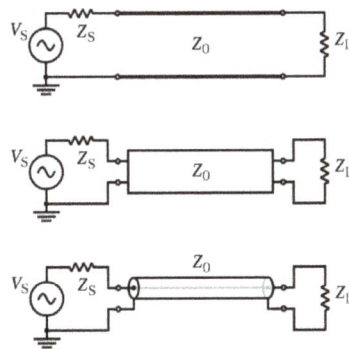

Variations on the schematic electronic symbol for a transmission line.

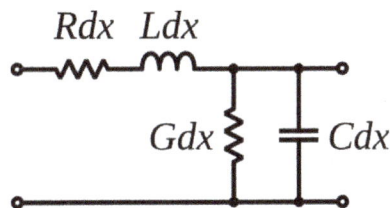

Schematic representation of the elementary components of a transmission line.

A transmission line is the material medium or structure that forms all or part of a path from one place to another for directing the transmission of energy, such as electromagnetic waves or electric power transmission. Types of transmission line include wires, coaxial cables, dielectric slabs, striplines, optical fibers, electric power lines and waveguides.

A microstrip is a type of transmission line that can be fabricated using printed circuit board technology and is used to convey microwave-frequency signals. It consists of a conducting strip separated from a ground plane by a dielectric layer known as the substrate. Microwave components such as antennas, couplers, filters and power dividers can be formed from a microstrip.

From the simplified schematics to the right it can be seen that total impedance, conductance, reactance (capacitance and inductance) and the transmission medium (transmission line) can be represented by single components that give the overall value.

With transmission line media it is important to match the load impedance Z_L to the characteristic impedance Z_0 as closely as possible, because it is usually desirable that the load absorbs as much power as possible.

      is the resistance per unit length,

      is the inductance per unit length,

      is the conductance of the dielectric per unit length,

      is the capacitance per unit length,

      is the imaginary unit, and

      is the angular frequency.

Lumped Circuit Elements

Often, because of the goal that moves physical metamaterial inclusions (or cells) to smaller sizes, discussion and implementation of lumped LC circuits or distributed LC networks are often examined. Lumped circuit elements are actually microscopic elements that effectively approximate their larger component counterparts. For example, circuit capacitance and inductance can be created with split rings, which are on the scale of nanometers at optical frequencies. The distributed LC model is related to the lumped LC model, however the distributed element model is more accurate but more complex than the lumped element model.

Metamaterial – Loaded Transmission Line Configurations

Some noted metamaterial antennas employ negative refractive index transmission-line metamaterials (NRI-TLM). These include lenses that can overcome the diffraction limit, small band and broadband phase shifting lines, small antennas, low profile antennas, antenna feed networks, novel power architectures and high directivity couplers. Loading a planar metamaterial network of TLs with series capacitors and shunt inductors produces higher performance. This results in a large operating bandwidth while the refractive index is negative.

Because superlenses can overcome the diffraction limit, this allows for a more efficient coupling to external radiation and enables a broader frequency band. For example, the superlens can be applied to the TLM architecture. In conventional lenses, imaging is limited by the diffraction limit.

With superlenses the details of the near field images are not lost. Growing evanescent waves are supported in the metamaterial ($n < 1$), which restores the decaying evanescent waves from the source. This results in a diffraction-limited resolution of $\lambda/6$, after some small losses. This compares with $\lambda/2$, the normal diffraction limit for conventional lenses.

By combining right-handed (RHM) with left-handed materials (LHM) as a composite material (CRLH) construction, both a backward to forward scanning capability is obtained.

Metamaterials were first used for antenna technology around 2005. This type of antenna used the established capability of SNGs to couple with external radiation. Resonant coupling allowed for a wavelength larger than the antenna. At microwave frequencies this allowed for a smaller antenna.

A metamaterial-loaded transmission line has significant advantages over conventional or standard delay transmission lines. It is more compact in size, it can achieve positive or negative phase shift while occupying the same short physical length and it exhibits a linear, flatter phase response with frequency, leading to shorter group delays. It can work in lower frequency because of high series distributed-capacitors and has smaller plane dimensions than its equivalent coplanar structure.

Negative Refractive Index Metamaterials Supporting 2-D Waves

In 2002, rather than using SRR-wire configuration, or other 3-D media, researchers looked at planar configurations that supported backward wave propagation, thus demonstrating negative refractive index and focusing as a consequence.

It has long been known that transmission lines periodically loaded with capacitive and inductive elements in a high-pass configuration support certain types of backward waves. In addition, planar transmission lines are a natural match for 2-D wave propagation. With lumped circuit elements they retain a compact configuration and can still support the lower RF range. With this in mind, high pass and cutoff, periodically loaded, two-dimensional LC transmission line networks were proposed. The LC networks can be designed to support backward waves, without bulky SRR/wire structure. This was the first such proposal which veered away from bulk media for a negative refractive effect. A notable property of this type of network is that there is no reliance on resonance, Instead the ability to support backward waves defines negative refraction.

The principles behind focusing are derived from Veselago and Pendry. Combining a conventional, flat, (planar) DPS slab, M-1, with a left-handed medium, M-2, a propagating electromagnetic wave with a wave vector k1 in M-1, results in a refracted wave with a wave vector k2 in M-2. Since, M-2 supports backward wave propagation k2 is refracted to the opposite side of the normal, while the Poynting vector of M-2 is anti-parallel with k2. Under such conditions, power is refracted through an effectively negative angle, which implies an effectively negative index of refraction.

Electromagnetic waves from a point source located inside a conventional DPS can be focused inside an LHM using a planar interface of the two media. These conditions can be modeled by exciting a single node inside the DPS and observing the magnitude and phase of the voltages to ground at all points in the LHM. A focusing effect should manifest itself as a "spot" distribution of voltage at a predictable location in the LHM.

Negative refraction and focusing can be accomplished without employing resonances or directly synthesizing the permittivity and permeability. In addition, this media can be practically fabricated by appropriately loading a host transmission line medium. Furthermore, the resulting planar topology permits LHM structures to be readily integrated with conventional planar microwave circuits and devices.

When transverse electromagnetic propagation occurs with a transmission line medium, the analogy for permittivity and permeability is $\varepsilon = L$, and $\mu = C$. This analogy was developed with positive values for these parameters. The next logic step was realizing that negative values could be achieved. In order to synthesize a left-handed medium ($\varepsilon < 0$ and $\mu < 0$) the series reactance and shunt susceptibility should become negative, because the material parameters are directly proportional to these circuit quantities.

A transmission line that has lumped circuit elements that synthesize a left-handed medium is referred to as a "dual transmission line" as compared to "conventional transmission line". The dual transmission line structure can be implemented in practice by loading a host transmission line with lumped element series capacitors (C) and shunt inductors (L). In this periodic structure, the loading is strong such that the lumped elements dominate the propagation characteristics.

Left-handed Behavior in LC Loaded Transmission Lines

Using SRRs at RF frequencies, as with wireless devices, requires the resonators to be scaled to larger dimensions. This worked against making the devices more compact. In contrast, LC network configurations could be scaled to both microwave and RF frequencies.

LC-loaded transmission lines enabled a new class of metamaterials to produce a negative refractive index. Relying on LC networks to emulate electrical permittivity and magnetic permeability resulted in a substantial increase in operating bandwidths.

Moreover, their unit cells are connected through a transmission-line network and may be equipped with lumped circuit elements, which permit them to be compact at frequencies where an SRR cannot be compact. The flexibility gained through the use of either discrete or printed elements enables planar metamaterials to be scalable from the megahertz to the tens of gigahertz range. In addition, replacing capacitors with varactors allowed the material properties to be dynamically tuned. The proposed media are planar and inherently support two-dimensional (2-D) wave propagation, making them well-suited for RF/microwave device and circuit applications.

Growing Evanescent Waves in Negative-refractive-index Transmission-line Media

The periodic 2-D LC loaded transmission-line (*TL*) was shown to exhibit NRI properties over a broad frequency range. This network will be referred to as a dual TL structure since it is of a high-pass configuration, as opposed to the low-pass representation of a conventional TL structure. Dual TL structures have been used to experimentally demonstrate backward-wave radiation and focusing at microwave frequencies.

As a negative refractive index medium, a dual TL structure is not simply a phase compensator. It can enhance the amplitude of evanescent waves, as well as correct the phase of propagating waves. Evanescent waves actually grow within the dual TL structure.

Backward Wave Antenna Using an NRI Loaded Transmission Line

Grbic *et al.* used one-dimensional LC loaded transmission line network, which supports fast backward-wave propagation to demonstrate characteristics analogous to "reversed Cherenkov radiation". Their proposed backward-wave radiating structure was inspired by negative refractive index LC materials. The simulated E-plane pattern at 15 GHz showed radiation towards the backfire direction in the far-field pattern, clearly indicating the excitation of a backward wave. Since the transverse dimension of the array is electrically short, the structure is backed by a long metallic trough. The trough acts as a waveguide below cut-off and recovers the back radiation, resulting in unidirectional far-field patterns.

Planar NIMs with Periodic Loaded Transmission Lines

Planar media can be implemented with an effective negative refractive index. The underlying concept is based on appropriately loading a printed network of transmission lines periodically with inductors and capacitors. This technique results in effective permittivity and permeability material parameters that are both inherently and simultaneously negative, obviating the need to employ separate means. The proposed media possess other desirable features including very wide bandwidth over which the refractive index remains negative, the ability to guide 2-D TM waves, scalability from RF to millimeter-wave frequencies and low transmission losses, as well as the potential for tunability by inserting varactors and/or switches in the unit cell. The concept has been verified with circuit and full-wave simulations. A prototype focusing device has been tested experimentally. The experimental results demonstrated focusing of an incident cylindrical wave within an octave bandwidth and over an electrically short area; suggestive of near-field focusing.

RF/microwave devices can be implemented based on these proposed media for applications in wireless communications, surveillance and radars.

Larger Transmission Lines

According to some researchers SRR/wire-configured metamaterials are bulky 3-D constructions that are difficult to adapt for RF/microwave device and circuit applications. These structures can achieve a negative index of refraction only within a narrow bandwidth. When applied to wireless devices at RF frequencies the split ring-resonators have to be scaled to larger dimensions, which, in turn forces a larger device size.

The proposed structures go beyond the wire/SRR composites in that they do not rely on SRRs to synthesize the material parameters, thus leading to dramatically increased operating bandwidths. Moreover, their unit cells are connected through a transmission-line network and they may, therefore, be equipped with lumped elements, which permit them to be compact at frequencies where the SRR cannot be compact. The flexibility gained through the use of either discrete or printed elements enables planar metamaterials to be scalable from the megahertz to the tens of gigahertz range. In addition, by utilizing varactors instead of capacitors, the effective material properties can be dynamically tuned. Furthermore, the proposed media are planar and inherently support two-dimensional (2-D) wave propagation. Therefore, these new metamaterials are well suited for RF/microwave device and circuit applications.

In the long-wavelength regime, the permittivity and permeability of conventional materials can be artificially synthesized using periodic LC networks arranged in a low-pass configuration. In the dual (high-pass) configuration, these equivalent material parameters assume simultaneously negative values, and may therefore be used to synthesize a negative refractive index.

Configurations

Antenna theory is based on classical electromagnetic theory as described by Maxwell's equations. Physically, an antenna is an arrangement of one or more conductors, usually called elements. An alternating current is created in the elements by applying a voltage at the antenna terminals, causing the elements to radiate an electromagnetic field. In reception, the reverse occurs: an electromagnetic field from another source induces an alternating current in the elements and a corresponding voltage at the antenna's terminals. Some receiving antennas (such as parabolic and horn types) incorporate shaped reflective surfaces to collect EM waves from free space and direct or focus them onto the actual conductive elements.

An antenna creates sufficiently strong electromagnetic fields at large distances. Reciprocally, it is sensitive to the electromagnetic fields impressed upon it externally. The actual coupling between a transmitting and receiving antenna is so small that amplifier circuits are required at both the transmitting and receiving stations. Antennas are usually created by modifying ordinary circuitry into transmission line configurations.

The required antenna for any given application is dependent on the bandwidth employed, and range (power) requirements. In the microwave to millimeter-wave range – wavelengths from a few meters to millimeters – the following antennas are usually employed:

Dipole antennas, short antennas, parabolic and other reflector antennas, horn antennas, periscope antennas, helical antennas, spiral antennas, surface-wave and leaky wave antennas. Leaky wave antennas include dielectric and dielectric loaded antennas, and the variety of microstrip antennas.

Radiation Properties with SRRs

The SRR was introduced by Pendry in 1999, and is one of the most common elements of metamaterials. As a nonmagnetic conducting unit, it comprises an array of units that yield an enhanced negative effective magnetic permeability, when the frequency of the incident electromagnetic field is close to the SRR resonance frequency. The resonant frequency of the SRR depends on its shape and physical design. In addition, resonance can occur at wavelengths much larger than its size.

Double Negative Metamaterials

Through the application of double negative metamaterials (DNG), the power radiated by electrically small dipole antennas can be notably increased. This could be accomplished by surrounding an antenna with a shell of double negative (DNG) material. When the electric dipole is embedded in a homogeneous DNG medium, the antenna acts inductively rather than capacitively, as it would in free space without the interaction of the DNG material. In addition, the dipole-DNG shell combination increases the real power radiated by more than an order of magnitude over a free space antenna. A notable decrease in the reactance of the dipole antenna corresponds to the increase in radiated power.

The reactive power indicates that the DNG shell acts as a natural matching network for the dipole. The DNG material matches the intrinsic reactance of this antenna system to free space, hence the impedance of DNG material matches free space. It provides a natural matching circuit to the antenna.

Single Negative SRR and Monopole Composite

The Addition Of An Srr-Dng Metamaterial increased the radiated power by more than an order of magnitude over a comparable free space antenna. Electrically small antennas, high directivity and tunable operational frequency are produced with negative magnetic permeability. When combining a right-handed material (RHM) with a Veselago-left-handed material (LHM) other novel properties are obtained. A single negative material resonator, obtained with an SRR, can produce an electrically small antenna when operating at microwave frequencies, as follows:

The configuration of an SRR assessed was two concentric annular rings with relative opposite gaps in the inner and outer ring. Its geometrical parameters were R = 3.6 mm, r = 2.5 mm, w = 0.2 mm, t = 0.9 mm. R and r are used in annular parameters, w is the spacing between the rings and t = the width of the outer ring. The material had a thickness of 1.6 mm. Permittivity was 3.85 at 4 GHz. The SRR was fabricated with an etching technique onto a 30 μm thick copper substrate. The SRR was excited by using a monopole antenna. The monopole antenna was composed of a coaxial cable, ground plane and radiating components. The ground plane material was aluminium. The operation frequency of the antenna was 3.52 GHz, which was determined by considering the geometrical parameters of SRR. An 8.32 mm length of wire was placed above the ground plane, connected to the antenna, which was one quarter of the operation wavelength. The antenna worked with a feed wavelength of 3.28 mm and feed frequency of 7.8 GHz. The SRR's resonant frequency was smaller than the monopole operation frequency.

The monopole-SRR antenna operated efficiently at (λ/10) using the SRR-wire configuration. It demonstrated good coupling efficiency and sufficient radiation efficiency. Its operation was comparable to a conventional antenna at λ/2, which is a conventional antenna size for efficient coupling and radiation. Therefore, the monopole-SRR antenna becomes an acceptable electrically small antenna at the SRR's resonance frequency.

When the SRR is made part of this configuration, characteristics such as the antenna's radiation pattern are entirely changed in comparison to a conventional monopole antenna. With modifications to the SRR structure the antenna size could reach (λ/40). Coupling 2, 3, and 4 SRRs side by side slightly shifts radiation patterns.

Patch Antennas

In 2005 a patch antenna with a metamaterial cover was proposed that enhanced directivity. According to the numerical results, the antenna showed significant improvement in directivity, compared to conventional patch antennae. This was cited in 2007 for an efficient design of directive patch antennas in mobile communications using metamaterials. This design was based on the left-handed material (LHM) transmission line model, with the circuit elements L and C of the LHM equivalent circuit model. This study developed formulae to determine the L and C values of the LHM equivalent circuit model for desirable characteristics of directive patch antennas. Design examples derived from actual frequency bands in mobile communications were performed, which illustrates the efficiency of this approach.

Flat Lens Horn Antenna

This configuration uses a flat aperture constructed of zero-index metamaterial. This has advantages over ordinary (conventional) curved lenses, which results in a much improved directivity. These investigations have provided capabilities for the miniaturization of microwave source and non-source devices, circuits, antennas and the improvement of electromagnetic performance.

Metamaterials Surface Antenna Technology

Metamaterials surface antenna technology (M-SAT) is an invention that uses metamaterials to direct and maintain a consistent broadband radio frequency beam locked on to a satellite whether the platform is in motion or stationary. Gimbals and motors are replaced by arrays of metamaterials in a planar configuration. Also, with this new technology phase shifters are not required as with phased array equipment. The desired affect is accomplished by varying the pattern of activated metamaterial elements as needed. The technology is a practical application of metamaterial cloaking theory. The antenna is approximately the size of a laptop computer.

Research and applications of metamaterial based antennas. Related components are also researched.

Subwavelength Cavities and Waveguides

When the interface between a pair of materials that function as optical transmission media interact as a result of opposing permittivity and / or permeability values that are either ordinary (positive) or extraordinary (negative), notable anomalous behaviors may occur. The pair would be a DNG metamaterial (layer), paired with a DPS, ENG or MNG layer. Wave propagation behavior and properties may occur that would otherwise not happen if only DNG layers are paired together.

At the interface between two media, the concept of the continuity of the tangential electric and magnetic field components can be applied. If either the permeability or permittivity of two media has opposite signs then the normal components of the tangential field, on both sides of the interface, will be discontinuous at the boundary. This implies a concentrated resonant phenomenon at the interface. This appears to be similar to the current and voltage distributions at the junction between an inductor and capacitor, at the resonance of an L-C circuit. This "*interface resonance*" is essentially independent of the total thickness of the paired layers, because it occurs along the discontinuity between two such conjugate materials.

Parallel-plate Waveguiding Structures

The geometry consists of two parallel plates as perfect conductors (PEC), an idealized structure, filled by two stacked planar slabs of homogeneous and isotropic materials with their respective constitutive parameters ε_1, ε_2, u_1, u_2. Each slab has thickness = d, slab 1 = d_1, and slab 2 = d_2.

Choosing which combination of parameters to employ involves pairing DPS and DNG or ENG and MNG materials. As mentioned previously, this is one pair of oppositely-signed constitutive parameters, combined.

Thin Subwavelength Cavity Resonators

Phase Compensation

The real component values for negative permittivity and permeability results in real component values for negative refraction n. In a lossless medium, all that would exist are real values. This concept can be used to map out phase compensation when a conventional lossless material, DPS, is matched with a lossless NIM (DNG).

In phase compensation, the DPS of thickness d_1 has $\varepsilon > 0$ and $\mu > 0$. Conversely, the NIM of thickness d_2 has $\varepsilon < 0$ and $\mu < 0$. Assume that the intrinsic impedance of the DPS dielectric material (d_1) is the same as that of the outside region and responding to a normally incident planar wave. The wave travels through the medium without any reflection because the DPS impedance and the outside impedance are equal. However, the plane wave at the end of DPS slab is out of phase with the plane wave at the beginning of the material.

The plane wave then enters the lossless NIM (d_2). At certain frequencies $\varepsilon < 0$ and $\mu < 0$ and n < 0. Like the DPS, the NIM has intrinsic impedance that is equal to the outside, and, therefore, is also lossless. The direction of power flow (i.e., the Poynting vector) in the first slab should be the same as that in the second one, because the power of the incident wave enters the first slab (without any reflection at the first interface), traverses the first slab, exits the second interface, enters the second slab and traverses it, and finally leaves the second slab. However, as stated earlier, the direction of power is anti-parallel to the direction of phase velocity. Therefore, the wave vector k_2 is in the opposite direction of k_1. Furthermore, whatever phase difference is developed by traversing the first slab can be decreased and even cancelled by traversing the second slab. If the ratio of the two thicknesses is $d_1 / d_2 = n_2 / n_1$, then the total phase difference between the front and back faces is zero. This demonstrates how the NIM slab at chosen frequencies acts as a phase compensator. It is important to note that this phase compensation process is only on the ratio of d_1 / d_2 rather than the thickness of $d_1 + d_1$. Therefore, $d_1 + d_1$ can be any value, as long as this ratio satisfies the above condition. Finally, even though this two-layer structure is present, the wave traversing this structure would not experience the phase difference.

Following this, the next step is the subwavelength cavity resonator.

Compact Subwavelength 1-D Cavity Resonators Using Metamaterials

The phase compensator described above can be used to conceptualize the possibility of designing a compact 1-D cavity resonator. The above two-layer structure is applied as two perfect reflectors, or in other words, two perfect conducting plates. Conceptually, what is constrained in the resonator is d_1 / d_2, not $d_1 + d_2$. Therefore, in principle, one can have a thin subwavelength cavity resonator for a given frequency, if at this frequency the second layer acts a metamaterial with negative permittivity and permeability and the ratio correlates to the correct values.

The cavity can conceptually be thin while still resonant, as long as the ratio of thicknesses is satisfied. This can, in principle, provide possibility for subwavelength, thin, compact cavity resonators.

Miniature Cavity Resonator Utilizing FSS

Frequency selective surface (FSS) based metamaterials utilize *equivalent* LC circuitry configurations. Using FSS in a cavity allows for miniaturization, decrease of the resonant frequency, lowers the cut-off frequency and smooth transition from a fast-wave to a slow-wave in a waveguide configuration.

Composite Metamaterial Based Cavities

As an LHM application four different cavities operating in the microwave regime were fabricated and experimentally observed and described.

Metamaterial Ground Plane

Leaky Mode Propagation with Metamaterial Ground Plane

A magnetic dipole was placed on metamaterial (slab) ground plane. The metamaterials have either constituent parameters that are both negative, or negative permittivity or negative permeability. The dispersion and radiation properties of leaky waves supported by these metamaterial slabs, respectively, were investigated.

Patented Systems

Phased array systems and antennas for use in such systems are well known in areas such as telecommunications and radar applications. In general phased array systems work by coherently re-assembling signals over the entire array by using circuit elements to compensate for relative phase differences and time delays.

Multiple systems have patents.

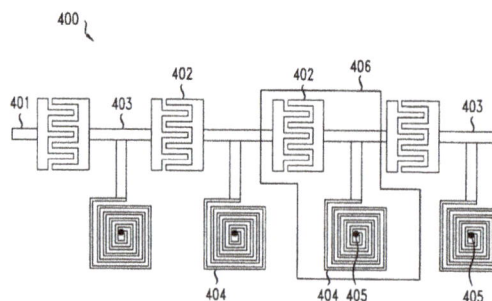

Microstrip line (400) for a phased array metamaterial antenna system. 401 represents unit-cellcircuits composed periodically along the microstrip. 402 series capacitors. 403 are T-junctions between capacitors,which connect (404) spiral inductor delay lines to 401. 404 are also connected to ground vias 405.

Phased Array Antenna

Patented in 2004, one phased array antenna system is useful in automotive radar applications. By using NIMs as a biconcave lens to focus microwaves, the antenna's sidelobes are reduced in size.

This equates to a reduction in radiated energy loss, and a relatively wider useful bandwidth. The system is an efficient, dynamically-ranged phased array radar system.

In addition, signal amplitude is increased across the microstrip transmission lines by suspending them above the ground plane at a predetermined distance. In other words, they are not in contact with a solid substrate. Dielectric signal loss is reduced significantly, reducing signal attenuation.

This system was designed to boost the performance of the Monolithic microwave integrated circuit (MMIC), among other benefits. A transmission line is created with photolithography. A metamaterial lens, consisting of a thin wire array focuses the transmitted or received signals between the line and the emitter / receiver elements.

The lens also functions as an input device and consists of a number of periodic unit-cells disposed along the line. The lens consists of multiple lines of the same make up; a plurality of periodic unit-cells. The periodic unit-cells are constructed of a plurality of electrical components; capacitors and inductors as components of multiple distributed circuits.

The metamaterial incorporates a conducting transmission element, a substrate comprising at least a first ground plane for grounding the transmission element, a plurality of unit-cell circuits composed periodically along the transmission element and at least one via for electrically connecting the transmission element to at least the first ground plane. It also includes a means for suspending this transmission element a predetermined distance from the substrate in a way such that the transmission element is located at a second predetermined distance from the ground plane.

ENG and MNG Waveguides and Scattering Devices

This structure was designed for use in waveguiding or scattering of waves. It employs two adjacent layers. The first layer is an epsilon-negative (ENG) material or a mu-negative (MNG) material. The second layer is either a double-positive (DPS) material or a double-negative (DNG) material. Alternatively, the second layer can be an ENG material when the first layer is an MNG material or the reverse.

Reducing Interference

A keyless entry system key fob

Metamaterials can reduce interference across multiple devices with smaller and simpler shielding. While conventional absorbers can be three inches thick, metamaterials can be in the millimeter range—2 mm (0.078 in) thick.

Reference

- Macnamara, Thereza (2010). Introduction to Antenna Placement and Installation. USA: John Wiley and Sons. p. 145. ISBN 0-470-01981-6.

- Straw, R. Dean (2003). The ARRL Antenna Book, 20th Edition. Newington, Connecticut, USA: The ARRL, Inc. p. 944. ISBN 0-87259-904-3.

- Silver, Ward (2007). The ARRL General Class License Manual, 6th Ed. USA: American Radio Relay League. p. 6.6. ISBN 978-0872599963.

- Tomasi, Wayne (2004). Electronic Communication Systems - Fundamentals Through Advanced. Jurong, Singapore: Pearson Education SE Asia Ltd. ISBN 981-247-093-X.

- Engheta, Nader; Richard W. Ziolkowski (June 2006). Metamaterials: physics and engineering explorations. Wiley & Sons. pp. 43–85. ISBN 978-0-471-76102-0.

- Iyer, Ashwin K.; George V. Eleftheriades (2002-06-07). "Negative Refractive Index Metamaterials Supporting 2-D Waves" (PDF). IEEE MTT-S International Microwave Symposium Digest. 2: 1067. doi:10.1109/MWSYM.2002.1011823. ISBN 0-7803-7239-5. Retrieved 2009-11-08.

- Bube, Richard H. (1992-09). Electrons in solids: an introductory survey. San Diego, CA: Elsevier Science. pp. 155, 156. ISBN 978-0-12-138553-8. Retrieved 2009-09-27.

- Fangming Zhu; Qingchun Lin; Jun Hu (2005). "2005 Asia-Pacific Microwave Conference Proceedings". 3: 1. doi:10.1109/APMC.2005.1606717. ISBN 0-7803-9433-X.

- Carr, Joseph J. (January 1998). "The Beverage Antenna". Popular Electronics. Farmington, IL: Gernsback Publications. 15 (1): 40–46. Retrieved July 1, 2016.

- Grbic, Anthony; George V. Eleftheriades (2003-03-24). "Growing evanescent waves in negative-refractive-index transmission-line media" (PDF). Applied Physics Letters. 82 (12): 1815. Bibcode:2003ApPhL..82.1815G. doi:10.1063/1.1561167. Archived from the original (PDF) on July 20, 2011. Retrieved 2009-11-30.

4

Essential Concepts of Antenna

Some of the theories stated are antenna farm, antenna aperture, antenna diversity and isotropic radiator. Antenna farm is the area dedicated to television and radio communications; usually an area with more than three antennas is referred to as an antenna farm. The chapter strategically encompasses and includes the essential components of antennas.

Antenna Diversity

Telephone exchange with later antennas mounted higher for antifade

Antenna diversity, also known as space diversity or spatial diversity, is any one of several wireless diversity schemes that uses two or more antennas to improve the quality and reliability of a wireless link. Often, especially in urban and indoor environments, there is no clear line-of-sight (LOS) between transmitter and receiver. Instead the signal is reflected along multiple paths before finally being received. Each of these bounces can introduce phase shifts, time delays, attenuations, and distortions that can destructively interfere with one another at the aperture of the receiving antenna.

Antenna diversity is especially effective at mitigating these multipath situations. This is because multiple antennas offer a receiver several observations of the same signal. Each antenna will experience a different interference environment. Thus, if one antenna is experiencing a deep fade, it is likely that another has a sufficient signal. Collectively such a system can provide a robust link. While this is primarily seen in receiving systems (diversity reception), the analog has also proven valuable for transmitting systems (transmit diversity) as well.

Inherently an antenna diversity scheme requires additional hardware and integration versus a single antenna system but due to the commonality of the signal paths a fair amount of circuitry can be shared. Also with the multiple signals there is a greater processing demand placed on the receiver, which can lead to tighter design requirements. Typically, however, signal reliability is paramount and using multiple antennas is an effective way to decrease the number of drop-outs and lost connections.

Antenna Techniques

Antenna diversity can be realized in several ways. Depending on the environment and the expected interference, designers can employ one or more of these methods to improve signal quality. In fact multiple methods are frequently used to further increase reliability.

- Spatial diversity employs multiple antennas, usually with the same characteristics, that are physically separated from one another. Depending upon the expected incidence of the incoming signal, sometimes a space on the order of a wavelength is sufficient. Other times much larger distances are needed. Cellularization or sectorization, for example, is a spatial diversity scheme that can have antennas or base stations miles apart. This is especially beneficial for the mobile communication industry since it allows multiple users to share a limited communication spectrum and avoid co-channel interference.

- Pattern diversity consists of two or more co-located antennas with different radiation patterns. This type of diversity makes use of directive antennas that are usually physically separated by some (often short) distance. Collectively they are capable of discriminating a large portion of angle space and can provide a higher gain versus a single omnidirectional radiator.

- Polarization diversity combines pairs of antennas with orthogonal polarizations (i.e. horizontal/vertical, ± slant 45°, Left-hand/Right-hand CP etc.). Reflected signals can undergo polarization changes depending on the medium through which they are travelling. A polarisation difference of 90° will result in an attenuation factor of up to 34 dB in signal strength. By pairing two complementary polarizations, this scheme can immunize a system from polarization mismatches that would otherwise cause signal fade. Additionally, such diversity has proven valuable at radio and mobile communication base stations since it is less susceptible to the near random orientations of transmitting antennas.

- Transmit/Receive diversity uses two separate, collocated antennas for transmit and receive functions. Such a configuration eliminates the need for a duplexer and can protect sensitive receiver components from the high power used in transmit.

- Adaptive arrays can be a single antenna with active elements or an array of similar antennas with ability to change their combined radiation pattern as different conditions persist. Active electronically scanned arrays (AESAs) manipulate phase shifters and attenuators at the face of each radiating site to provide a near instantaneous scan ability as well as pattern and polarization control. This is especially beneficial for radar applications since it affords a single antenna the ability to switch among several different modes such as searching, tracking, mapping and jamming countermeasures.

Processing Techniques

All of the above techniques require some sort of post processing to recover the desired message. Among these techniques are:

- Switching – In a switching receiver, the signal from only one antenna is fed to the receiver for as long as the quality of that signal remains above some prescribed threshold. If and when the signal degrades, another antenna is switched in. Switching is the easiest and least power consuming of the antenna diversity processing techniques but periods of fading and desynchronization may occur while the quality of one antenna degrades and another antenna link is established.

- Selecting – As with switching, selection processing presents only one antenna's signal to the receiver at any given time. The antenna chosen, however, is based on the best signal-to-noise ratio (SNR) among the received signals. This requires that a pre-measurement take place and that all antennas have established connections (at least during the SNR measurement) leading to a higher power requirement. The actual selection process can take place in between received packets of information. This ensures that a single antenna connection is maintained as much as possible. Switching can then take place on a packet-by-packet basis if necessary.

- Combining – In combining, all antennas maintain established connections at all times. The signals are then combined and presented to the receiver. Depending on the sophistication of the system, the signals can be added directly (equal gain combining) or weighted and added coherently (maximal-ratio combining). Such a system provides the greatest resistance to fading but since all the receive paths must remain energized, it also consumes the most power.

- Dynamic Control – Dynamically controlled receivers are capable of choosing from the above processing schemes for whenever the situation arises. While much more complex, they optimize the power vs. performance trade-off. Transitions between modes and/or antenna connections are signaled by a change in the perceived quality of the link. In situations of low fading, the receiver can employ no diversity and use the signal presented by a single antenna. As conditions degrade, the receiver can then assume the more highly reliable but power-hungry modes described above.

Applications

A well-known practical application of diversity reception is in wireless microphones, and in similar electronic devices such as wireless guitar systems. A wireless microphone with a non-diversity receiver (a receiver having only one antenna) is prone to random drop-outs, fades, noise, or other interference, especially if the transmitter (the wireless microphone) is in motion. A wireless microphone or sound system using diversity reception will switch to the other antenna within microseconds if one antenna experiences noise, providing an improved quality signal with fewer drop-outs and noise. Ideally, no drop-outs or noise will occur in the received signal.

Another common usage is in Wi-Fi networking gear and cordless telephones to compensate for multipath interference. The base station will switch reception to one of two antennas depending on which is currently receiving a stronger signal. For best results, the antennas are usually placed one wavelength apart. For microwave bands, where the wavelengths are under 100 cm, this can often be done with two antennas attached to the same hardware. For lower frequencies and longer wavelengths, the antennas must be several meters apart, making it much less reasonable.

Mobile phone towers also often take advantage of diversity - each face (sector) of a tower will often have two antennas; one is transmitting and receiving, while the other is a receive only antenna. Two receivers are used to perform diversity reception.

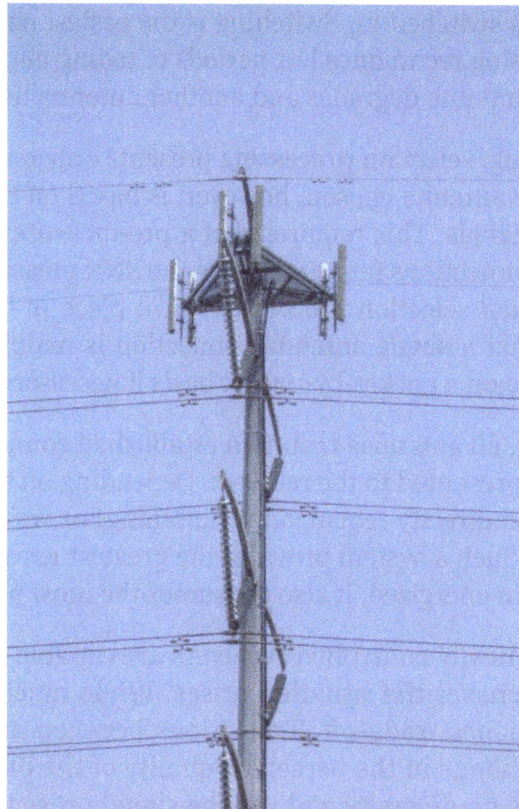

Cell antennas on an electricity pylon showing two antennas per sector

The use of multiple antennas at both transmit and receive results in a multiple-input multiple-output (MIMO) system. The use of diversity techniques at both ends of the link is termed space–time coding.

Antenna Diversity for MIMO

Diversity Coding is the spatial coding techniques for a MIMO system in wireless channels. Wireless channels severely suffer from fading phenomena, which causes unreliability in data decoding. Fundamentally, diversity coding sends multiple copies through multiple transmit antennas, so as to improve the reliability of the data reception. If one of them fails to receive, the others are used for data decoding. MIMO achieves spatial diversity and spatial multiplexing.

Antenna Aperture

In electromagnetics and antenna theory, antenna aperture or effective area is a measure of how effective an antenna is at receiving the power of radio waves. The aperture is defined as the area, oriented perpendicular to the direction of an incoming radio wave, which would intercept the same amount of power from that wave as is produced by the antenna receiving it. At any point, a beam of radio waves has an *irradiance* or *power flux density (PFD)* which is the amount of radio power passing through a unit area of one square meter. If an antenna delivers an output power of P_o watts to the load connected to its output terminals when irradiated by a uniform field of power density *PFD* watts per square metre, the antenna's aperture A_{eff} in square metres is given by:

$$A_{eff} = \frac{P_o}{PFD}.$$

So the power output of an antenna in watts is equal to the power density of the radio waves in watts per square metre, multiplied by its aperture in square metres. The larger an antenna's aperture is, the more power it can collect from a given field of radio waves. To actually obtain the predicted power available P_o, the polarization of the incoming waves must match the polarization of the antenna, and the load (receiver) must be impedance matched to the antenna's feedpoint impedance.

Although this concept is based on an antenna receiving a radio frequency wave, knowing A_{eff} directly supplies the (power) gain of that antenna. Due to reciprocity, an antenna's gain in receiving and transmitting are identical. Therefore A_{eff} can just as well be used to compute the performance of a transmitting antenna. Note that A_{eff} is a function of the direction of the radio wave relative to the orientation of the antenna, since the gain of an antenna varies according to its radiation pattern. When no direction is specified, A_{eff} is understood to refer to its maximum value, with the antenna oriented so its main lobe, the axis of maximum sensitivity, is directed toward the source.

Aperture Efficiency

In general, the aperture of an antenna is not directly related to its physical size. However some types of antennas, for example parabolic dishes and horns, have a physical aperture (opening) which collects the radio waves. In these *aperture antennas*, the effective aperture A_{eff} must always be less than the area of the antenna's physical aperture A_{phys}, as can be seen from the definition above. An antenna's *aperture efficiency*, e_a is defined as the ratio of these two areas:

$$e_a = \frac{A_{eff}}{A_{phys}}$$

The aperture efficiency is a dimensionless parameter between 0 and 1.0 that measures how close the antenna comes to using all the radio power entering its physical aperture. If the antenna were perfectly efficient, all the radio power falling within its physical aperture would be converted to electrical power delivered to the load attached to its output terminals, so these two areas would be equal $A_{eff} = A_{phys}$ and the aperture efficiency would be 1.0. But all antennas have losses, such as power dissipated as heat in the resistance of its elements, nonuniform illumination by its feed, and radio waves scattered by structural supports and diffraction at the aperture edge, which reduce the power output. Aperture efficiencies of typical antennas vary from 0.35 to 0.70 but can range up to 0.90.

Aperture and Gain

The directivity of an antenna, its ability to direct radio waves in one direction or receive from a single direction, is measured by a parameter called its gain, which is the ratio of the power received by the antenna to the power that would be received by a hypothetical isotropic antenna, which receives power equally well from all directions.

As shown in the article on isotropic radiators, the aperture of a lossless isotropic antenna, which by definition has unity gain, is:

$$A_{eff} = \frac{\lambda^2}{4\pi}$$

where λ is the wavelength of the radio waves. So the gain of any antenna is proportional to its aperture:

$$G = \frac{A_{eff}}{\lambda^2 / 4\pi} = \frac{4\pi A_{phys} e_a}{\lambda^2}$$

So antennas with large effective apertures are high gain antennas, which have small angular beam widths. Most of their power is radiated in a narrow beam in one direction, and little in other directions. As receiving antennas, they are most sensitive to radio waves coming from one direction, and are much less sensitive to waves coming from other directions. Although these terms can be used as a function of direction, when no direction is specified, the gain and aperture are understood to refer to the antenna's axis of maximum gain, or boresight.

Friis Transmission Equation

The fraction of the power delivered to a transmitting antenna that is received by a receiving antenna is proportional to the product of the apertures of both the antennas. This is given by a form of the Friis transmission equation:.

$$P_r = \frac{A_t A_r}{r^2 \lambda^2} P_t$$

where

P_r is the power delivered by the receiving antenna in watts

P_t is the power applied to the transmitting antenna in watts

A_r is the aperture of the receiving antenna in m²

A_t is the aperture of the transmitting antenna in m²

r is the distance between the antenna in m

λ is the wavelength of the radio waves in m

Thin Element Antennas

In the case of thin element antennas such as monopoles and dipoles, there is no simple relationship between physical area and effective area. However, the effective areas can be calculated from their power gain figures:

Wire antenna	Power gain	Effective area
Short dipole (Hertzian dipole)	1.5	0.1194 ^{2}
Half-wave dipole	1.64	0.1305 ^{2}
Quarter-wave mon opole	3.28	0.2610 ^{2}

This assumes that the monopole antenna is mounted above an infinite ground plane and that the antennas are lossless. When resistive losses are taken into account, particularly with small antennas, the antenna gain might be substantially less than the directivity, and the effective area is less by the same factor.

Effective Length

For antennas which are not defined by a physical area, such as monopoles and dipoles consisting of thin rod conductors, the aperture bears no obvious relation to the size or area of the antenna. An alternate measure of antenna gain that has a greater relationship to the physical structure of such antennas is *effective length* l_{eff} measured in metres, which is defined for a receiving antenna as:

$$l_{eff} = V_0 >$$

where

V_o is the open circuit voltage appearing across the antenna's terminals

E_s is the electric field strength of the radio signal, in volts per metre, at the antenna.

The longer the effective length the more voltage and therefore the more power the antenna will receive. Note, however, that an antenna's gain or A_{eff} increases according to the *square* of l_{eff}, and that this proportionality also involves the antenna's radiation resistance. Therefore this measure is of more theoretical than practical value and is not, by itself, a useful figure of merit relating to an antenna's directivity.

Antenna Factor

In electromagnetics, the antenna factor is defined as the ratio of the electric field strength to the voltage V (units: V or µV) induced across the terminals of an antenna. The voltage measured at the output terminals of an antenna is not the actual field intensity due to actual antenna gain, aperture characteristics, and loading effects.

For an electric field antenna, the field strength is in units of V/m or µV/m and the resulting antenna factor AF is in units of 1/m:

$$AF = \frac{E}{V}$$

If all quantities are expressed logarithmically in decibels instead of SI units, the above equation becomes

$$AF_{\text{dBm}^{-1}} = E_{\text{dBV/m}} - V_{\text{dBV}} = E_{\text{dB}\mu\text{V/m}} - V_{\text{dB}\mu\text{V}}$$

For a magnetic field antenna, the field strength is in units of A/m and the resulting antenna factor is in units of A/(Vm). For the relationship between the electric and magnetic fields, see the impedance of free space.

For a 50 Ω load, knowing that $P_D A_e = P_r = V^2/R$ and $E^2 = 377P_D$, the antenna factor is developed as:

$$AF = \frac{\sqrt{377P_D}}{\sqrt{50P_D A_e}} = \frac{2.75}{\sqrt{A_e}} = \frac{9.73}{\lambda\sqrt{G}}$$

Where

- $A_e = (\lambda^2 G)/4\pi$: the antenna effective aperture

- G: the antenna gain

For antennas which are not defined by a physical area, such as monopoles and dipoles consisting of thin rod conductors, the effective length is used to measure the ratio between E and V.

Antenna Gain

In electromagnetics, an antenna's power gain or simply gain is a key performance number which combines the antenna's directivity and electrical efficiency. As a transmitting antenna, the gain describes how well the antenna converts input power into radio waves headed in a specified direction. As a receiving antenna, the gain describes how well the antenna converts radio waves arriving from a specified direction into electrical power. When no direction is specified, "gain" is understood to refer to the peak value of the gain. A plot of the gain as a function of direction is called the radiation pattern.

Antenna gain is usually defined as the ratio of the power produced by the antenna from a far-field source on the antenna's beam axis to the power produced by a hypothetical lossless *isotropic antenna*, which is equally sensitive to signals from all directions. Usually this ratio is expressed in decibels, and these units are referred to as "*decibels-isotropic*" (dBi). An alternative definition compares the antenna to the power received by a lossless half-wave dipole antenna, in which case the units are written as *dBd*. Since a lossless dipole antenna has a gain of 2.15 dBi, the relation between these units is: *gain in dBd = gain in dBi – 2.15*. For a given frequency, the antenna's effective area is proportional to the power gain. An antenna's effective length is proportional to the *square root* of the antenna's gain for a particular frequency and radiation resistance. Due to reciprocity, the gain of any antenna when receiving is equal to its gain when transmitting.

Directive gain or directivity is a different measure which does *not* take an antenna's electrical efficiency into account. This term is sometimes more relevant in the case of a receiving antenna where one is concerned mainly with the ability of an antenna to receive signals from one direction while rejecting interfering signals coming from a different direction.

Power Gain

Power gain (or simply gain) is a unitless measure that combines an antenna's efficiency $E_{antenna}$ and directivity D:

$$G = E_{antenna} \cdot D.$$

The notions of efficiency and directivity depend on the following.

EfficiencyA transmitting antenna accepts input power P_{in} at some point along the feedline. The point is typically taken to be at the antenna (the *feedpoint*), thereby not counting power lost due to joule heating in the feedline and reflections back down the feedline. An antenna with efficiency $E_{antenna}$ emits a total radiated power

$$P_o = E_{antenna} \cdot P_{in}$$

to its environment. The environment may range from free space (the default when not specified) to an object such as a hand surrounding the antenna. Reciprocity justifies taking the properties of a receiving antenna, such as efficiency, directivity, and gain, to be those of that antenna when used for transmission.

Directivity

Antennas are invariably directional to a greater or lesser extent, according to how the output power is distributed in any given direction in three dimensions. We shall specify direction here in spherical coordinates (θ, ϕ), where θ is the altitude or angle above a specified reference plane (such as the ground), while ϕ is the azimuth as the angle between the projection of the given direction onto the reference plane and a specified reference direction (such as north or east) in that plane with specified sign (either clockwise or counterclockwise).

The distribution of output power as a function of the possible directions (θ, ϕ) is given by its radiation intensity $U(\theta, \phi)$ (in SI units: watts per steradian, Wsr^{-1}). The output power is obtained from the radiation intensity by integrating the latter over all directions:

$$P_o = \int_{-\pi}^{\pi}\int_{-\pi/2}^{\pi/2} U(\theta,\phi)d\Omega = \int_{-\pi}^{\pi}\int_{-\pi/2}^{\pi/2} U(\theta,\phi)\cos\theta d\theta d\phi.$$

The mean radiation intensity $\bar{U}$ is therefore given by

$$\bar{U} = \frac{P_o}{4\pi}$$ since there are 4π steradians in a sphere

$$= \frac{E_{antenna} \cdot P_{in}}{4\pi} \text{ using the first formula for } P_o.$$

The directive gain or directivity $D(\theta,\phi)$ of an antenna in a given direction is the ratio of its radiation intensity $U(\theta,\phi)$ in that direction to its mean radiation intensity $U(\theta,\phi)$. That is,

$$D(\theta,\phi) = \frac{U(\theta,\phi)}{\bar{U}}.$$

An isotropic antenna, meaning one with the same radiation intensity in all directions, therefore has directivity 1 in all directions independently of its efficiency. More generally the maximum, minimum, and mean directivities of any antenna are always at least 1, at most 1, and exactly 1. For the half-wave dipole the respective values are 1.64 (2.15 dB), 0, and 1.

When the directivity D of an antenna is given independently of direction it refers to its maximum directivity in any direction, namely

$$D = \max_{\theta,\phi} D(\theta,\phi)$$

Gain

The power gain or simply gain $G(\theta,\phi)$. of an antenna in a given direction takes efficiency into account by being defined as the ratio of its radiation intensity $U(\theta,\phi)$ in that direction to the mean radiation intensity of a perfectly efficient antenna. Since the latter equals $P_{in}/4\pi$, it is therefore given by

$$G(\theta,\phi) = \frac{U(\theta,\phi)}{P_{in}/4\pi}$$

$$= E_{antenna} \cdot \frac{U(\theta,\phi)}{\bar{U}} \text{ using the second equation for } \bar{U}$$

$$= E_{antenna} \cdot D(\theta,\phi) \text{ using the equation for } D(\theta,\phi).$$

As with directivity, when the gain G of an antenna is given independently of direction it refers to its maximum gain in any direction. Since the only difference between gain and directivity in any directio :

$$G = E_{antenna} \cdot D.$$

Summary

If only a certain portion of the electrical power received from the transmitter is actually radiated by the antenna (i.e. less than 100% efficiency), then the directive gain compares the power radiated in a given direction to that reduced power (instead of the total power received), ignoring the inefficiency. The directivity is therefore the maximum directive gain when taken over all directions, and is always *at least* 1. On the other hand, the power gain takes into account the poorer efficiency by comparing the radiated power in a given direction to the actual power that the antenna receives from the transmitter, which makes it a more useful figure of merit for the antenna's contribution to the ability of a transmitter in sending a radio wave toward a receiver. In every direction, the power gain of an isotropic antenna is equal to the efficiency, and hence is always *at most* 1, though it can and ideally should exceed 1 for a directional antenna.

Note that in the case of an impedance mismatch, P_{in} would be computed as the transmission line's incident power minus reflected power. Or equivalently, in terms of the rms voltage V at the antenna terminals:

$$P_{in} = V^2 \cdot Re\left\{\frac{1}{Z_{in}}\right\}$$

where Z_{in} is the feedpoint impedance.

Gain in Decibels

Published numbers for antenna gain are almost always expressed in decibels (dB), a logarithmic scale. From the gain factor G, one finds the gain in decibels as:

$$G_{dBi} = 10 \cdot \log_{10}(G)$$

Therefore, an antenna with a peak power gain of 5 would be said to have a gain of 7 dBi. "dBi" is used rather than just "dB" to emphasize that this is the gain according to the basic definition, in which the antenna is compared to an isotropic radiator.

When actual measurements of an antenna's gain are made by a laboratory, the field strength of the test antenna is measured when supplied with, say, 1 watt of transmitter power, at a certain distance. That field strength is compared to the field strength found using a so-called *reference antenna* at the same distance receiving the same power in order to determine the gain of the antenna under test. That ratio would be equal to G if the reference antenna were an isotropic radiator.

However a true isotropic radiator cannot be built, so in practice a different antenna is used. This will often be a half-wave dipole, a very well understood and repeatable antenna that can be easily built for any frequency. The directive gain of a half-wave dipole is known to be 1.64 and it can be made nearly 100% efficient. Since the gain has been measured with respect to this reference antenna, the difference in the gain of the test antenna is often compared to that of the dipole. The "gain relative to a dipole" is thus often quoted and is denoted using "dBd" instead of "dBi" to avoid confusion. Therefore, in terms of the true gain (relative to an isotropic radiator) G, this figure for the gain is given by:

$$G_{dBd} = 10 \cdot \log_{10}\left(\frac{G}{1.64}\right)$$

For instance, the above antenna with a gain G=5 would have a gain with respect to a dipole of 5/1.64 = 3.05, or in decibels one would call this 10 log(3.05) = 4.84 dBd. In general:

$$G_{dBd} = G_{dBi} - 2.15dB$$

Both dBi and dBd are in common use. When an antenna's maximum gain is specified in decibels (for instance, by a manufacturer) one must be certain as to whether this means the gain relative to an isotropic radiator or with respect to a dipole. If it specifies "dBi" or "dBd" then there is no ambiguity, but if only "dB" is specified then the fine print must be consulted. Either figure can be easily converted into the other using the above relationship.

Note that when considering an antenna's directional pattern, "gain with respect to a dipole" does *not* imply a comparison of that antenna's gain in each direction to a dipole's gain in that direction. Rather, it is a comparison between the antenna's gain in each direction to the *peak* gain of the dipole (1.64). In any direction, therefore, such numbers are 2.15 dB smaller than the gain expressed in dBi.

Partial Gain

Partial gain is calculated as power gain, but for a particular polarization. It is defined as the part of the radiation intensity U corresponding to a given polarization, divided by the total radiation intensity of an isotropic antenna.

$$G_\theta = 4\pi \left(\frac{U_\theta}{P_{in}} \right)$$

$$G_\phi = 4\pi \left(\frac{U_\phi}{P_{in}} \right)$$

where U_θ and U_ϕ represent the radiation intensity in a given direction contained in their respective E field component.

As a result of this definition, we can conclude that the total gain of an antenna is the sum of partial gains for any two orthogonal polarizations.

$$G = G_\theta + G_\phi$$

Example Calculation

Suppose a lossless antenna has a radiation pattern given by:

$$U = B_0 \sin^3(\theta)$$

Let us find the gain of such an antenna.

Solution:

First we find the peak radiation intensity of this antenna:

$$U_{max} = B_0$$

The total radiated power can be found by integrating over all directions:

$$P_{rad} = \int_0^{2\pi} \int_0^{\pi} U(\theta,\phi)\sin(\theta)d\theta d\phi = 2\pi B_0 \int_0^{\pi} \sin^4(\theta)d\theta = B_0\left(\frac{3\pi^2}{4}\right)$$

$$D = 4\pi\left(\frac{U_{max}}{P_{rad}}\right) = 4\pi\left[\frac{B_0}{B_0\left(\frac{3\pi^2}{4}\right)}\right] = \frac{16}{3\pi} = 1.698$$

Since the antenna is specified as being lossless the radiation efficiency is

1. The maximum gain is then equal

$$G = E_{antenna}D = (1)(1.698) = 1.698. \cdot$$
$$G_{dBi} = 10\log_{10}(1.698) = 2.30 dBi$$

Expressed relative to the gain of a half-wave dipole we would find:

$$G_{dBd} = 10\log_{10}(1.698/1.64) = 0.15 dBd. \cdot$$

Realized Gain

According to IEEE Standard 145-1993, Realized Gain differs from the above definitions of gain in that it is "reduced by the losses due to the mismatch of the antenna input impedance to a specified impedance." This mismatch induces losses above the dissipative losses described above; therefore, Realized Gain will always be less than Gain.

Gain may be expressed as absolute gain if further clarification is required to differentiate it from Realized Gain.

Total Radiated Power

(TRP)Total radiated power is the sum of all RF power radiated by the antenna when the source power is included in the measurement. TRP is expressed in Watts, or equivalent logarithmic expressions, often dBm or dBW.

TRP can be measured while in the close proximity of power-absorbing losses such as the body and hand of the Mobile Device Under Test User.

The TRP can be used to determine Body Loss (BoL). The Body Loss is considered as the ratio of TRP measured in the presence of losses and TRP measured while in free space.

Antenna Height Considerations

The Aspects for Antenna heights considerations are depending upon the wave range and economical reasons.

Longwave/Low Frequency Antennas

At VLF, LF and MF the radio mast or tower is often used directly as an antenna. Its height determines the vertical radiation pattern. Masts and towers with heights around a quarter wave or shorter, radiate considerable power towards the sky. This allows only a small area of fade-free reception at night, because the distance at which groundwave and skywave are of comparable strength and can interfere with each other is severely restricted (approximately 40 kilometres to 200 kilometres from the transmission site, depending on frequency and ground conductivity).

For high power transmitters, masts with heights of about half the radiated wavelength are preferred because they concentrate the radiated power toward the horizon. This enlarges the distance at which selective fading occurs. However, masts with heights of around half a wavelength are much more expensive than shorter ones and often too expensive for lower power mediumwave stations.

For longwave transmitters, however, the construction of halfwave masts is infeasible in most cases, either for economic reasons or because of problems with flight safety. The only radio mast for longwave with a height of half a wavelength built to date was the Warsaw Radio Mast (which did not survive). For frequencies lower than the longwave range, masts have to be electrically enlarged by base loading coils or structures on the top, because the heights required for masts of even a quarter wavelength are too large to realize physically.

Use is not normally made of masts higher than five-eighths (5/8) of a wavelength, because such masts (except for some special constructions for high power mediumwave broadcasting) exhibit poor vertical radiation patterns. The heights of masts for mediumwave transmitters normally do not exceed the 300 metre (1000 foot) level.

Sometimes cage aerials or longwire aerials are used for LF and MF transmission. In this case the height of the tower may be greater than is usually the case. Because towers or masts used for cage or long wire aerials are grounded at the base, they are especially suitable for supporting antennas for UHF or VHF broadcasting.

Shortwave/High Frequency Antennas

For transmissions in the shortwave range, mast height has no influence on efficiency. Masts are generally used to support the antenna. Most shortwave masts are less than 100 metres high.

Antennas for Commercial UHF/VHF

for transmissions in the VHF and UHF range, tower importance and value can vary depending on the area to be served. The cost of a tower must be recouped primarily through advertising carried on the broadcasts, especially where there are no license fees charged the listener. Considerations such as population density, line-of-sight signal range (affected by terrain), and the costs of tower construction and maintenance versus height, must all be weighed in choosing an ideal tower size. Often there are restrictions related to flight safety governing maximum allowable tower height. Two shorter towers may be a better option then a single taller tower. Also, a higher tower might not be useful if the signal is blocked by terrain or if all the listeners are in a concentrated area and a higher tower cannot pay for itself.

Antennas for Point-to-point Radio Services

In most applications line-of-sight is required between the transmitting and receiving antennas for point-to-point services, and antennas may have to be mounted at a certain height about ground. For microwave radio systems, it is not always possible to use long transmission lines between transmitter and antenna, so towers with equipment rooms or cabinets close to the height of the antenna may be required.

Radio Propagation Beacon

A radio propagation beacon is a radio beacon, whose purpose is the investigation of the propagation of radio signals. Most radio propagation beacons use amateur radio frequencies. They can be found on LF, MF, HF, VHF, UHF, and microwave frequencies. Microwave beacons are also used as signal sources to test and calibrate antennas and receivers.

The International Amateur Radio Union (IARU) and its member societies coordinate beacons established by radio amateurs.

Propagation beacon 4U1UN, transmitting from the UN Building in New York.

Transmission Characteristics

Most beacons operate in continuous wave (A1A) and transmit their identification (call sign and location). Some of them send long dashes to facilitate signal strength measurement. A small number of beacons transmit Morse code by frequency shift keying (F1A). A few beacons transmit signals in digital modulation modes, like radioteletype (F1B) and PSK31 (G1B).

2200 Meter Beacons

Amateur experiments in the 2200-meter band (135.7-137.8 kHz) often involve operating temporary beacons.

1750 Meter Beacons

In the United States and Canada, unlicensed experimenters called Lowfers establish low power beacons on radio frequencies between 160 kHz and 190 kHz.

160 Meter Beacons

The International Amateur Radio Union Region 2 (North and South America) bandplan for 160 meters reserves the range 1999 kHz to 2000 kHz for propagation beacons.

10 Meter Beacons

Most high frequency radio propagation beacons are found in the 10 meters (28 MHz) frequency band, where they are good indicators of Sporadic E ionospheric propagation. According to IARU bandplans, the following 28 MHz frequencies are allocated to radio propagation beacons:

IARU Region	Beacon Sub-bands
R1	• 28190-28199 Regional Time Shared • 28199-28201 The International Beacon Project • 28201-28225 Continuous Duty
R2	• 28190-28199 Regional Time Shared • 28199-28201 The International Beacon Project • 28201-28225 Beacons, continuous duty • 28225-28300 Shared
R3	• 28190-28200 IBP

6 Meter Beacons

In the 6 meter (50 MHz) band, beacons operate in the lower part of the band, traditionally in the range 50.000 MHz to 50.080 MHz. The IARU is encouraging individual beacons to move to 50.4 MHz to 50.5 MHz to assist with the establishment of the *Synchronised 50 MHz Beacon Project*. In the United States, the Federal Communications Commission (FCC) only permits unattended 6 meter beacon stations to operate between 50.060 and 50.080 MHz. Due to unpredictable and intermittent long distance propagation, usually achieved by a combination of ionospheric conditions, beacons are very important in providing early warning for 50 MHz openings.

4 Meter Beacons

Several countries in ITU Region 1 have access to frequencies in the 70 MHz region, called the 4 meters band. The band shares many propagation characteristics with 6 meters. The preferred location for beacons is 70,000 to 70,090 kHz; however, in countries where this segment is not allocated to Amateur Radio, beacons may operate elsewhere in the band.

VHF/UHF Beacons

Beacons on 144 MHz and higher frequencies are mainly used to identify tropospheric radio propagation openings. It is not uncommon for VHF and UHF beacons to use directional antennas. Frequencies set aside for beacons on VHF and UHF bands vary widely in different ITU regions and countries.

Band	Beacon Sub-band (MHz)		
	ITU Region 1	ITU Region 2	ITU Region 3
2 m	144.400-144.491	144.275–144.300	Unknown
1.25 m	N/A	222.050–222.060	N/A
70 cm	432.400-432.490	432.300–432.400	Unknown
33 cm	N/A	903.000-903.100	N/A
23 cm	1,296.800-1,296.994	1,296.200-1,296.400	Unknown
13 cm	2,320.800-2,321.000	2,304.300-2,304.400	Unknown

The beacon sub-bands in the United Kingdom, also reflect IARU Region 1 recommendations.

SHF/Microwave Beacons

In addition to identifying propagation, microwave beacons are also used as signal sources to test and calibrate antennas and receivers. SHF beacons are not as common as beacons on the lower bands, and beacons above the 3 centimeters band (10 GHz) are unusual.

Band	Beacon Sub-band (MHz)		
	ITU Region 1	ITU Region 2	ITU Region 3
9 cm	3,400.800-3,400.995	3,456.300-3,457.000	Unknown
5 cm	5,760.800-5,760.990	5,760.300-5,760.400	Unknown
3 cm	10,368.800-10,368.990	10,368.300-10,368.400	Unknown
1.2 cm	24,048.800-24,048.995	24,048.800-24,048.995	Unknown

Beacon Projects

Most radio propagation beacons are operated by individual radio amateurs or amateur radio societies and clubs. As a result, there are frequent additions and deletions to the lists of beacons. There are, however a few major projects coordinated by organizations like the International Amateur Radio Union.

IARU Beacon Project

Beacons from Finland and Madeira on 14.100 MHz

The International Beacon Project (IBP), which is coordinated by the Northern California DX Foundation and the International Amateur Radio Union, consists of 18 high frequency propagation beacons worldwide, which transmit in turns on 14.100 MHz, 18.110 MHz, 21.150 MHz, 24.930 MHz, and 28.200 MHz.

DARC Beacon Project

The Deutscher Amateur-Radio-Club sponsors two beacons which transmit from Scheggerott, near Kiel (JO44vq). These beacons are DRA5 on 5195 kHz and DK0WCY on 10144 kHz. In addition to identification and location, every 10 minutes these beacons transmit solar and geomagnetic bulletins. Transmissions are in Morse code for aural reception, RTTY and PSK31. DK0WCY operates also a limited service beacon on 3579 kHz at 0720–0900 and 1600–1900 local time.

RSGB 5 MHz Beacon Project

The Radio Society of Great Britain operates three radio propagation beacons on 5290 kHz, which transmit in sequence, for one minute each, every 15 minutes. The project includes GB3RAL near Didcot (IO91in), GB3WES in Cumbria (IO84qn) and GB3ORK in the Orkney Islands (IO89ja). GB3RAL, which is located at the Rutherford-Appleton Laboratory, also transmits continuously on 28215 kHz and on a number of low VHF frequencies (40050, 50050, 60050 and 70050 kHz).

Weak Signal Propagation Reporter Network

A large-scale amateur radio propagation beacon project is underway using the WSPR transmission scheme included with the WSJT software suite. The loosely coordinated beacon transmitters and receivers, collectively known as the WSPRnet, report the real-time propagation characteristics of a number of frequency bands and geographical locations via the Internet. The WSPRnet website provides detailed propagation report databases and real-time graphical maps of propagation paths.

Past Beacon Projects

As part of an International Telecommunications Union-funded project, radio propagation beacons were installed by national authorities at Sveio, Norway (callsign LN2A, JO29po) and at Darwin, Australia (callsign VL8IPS, PH57pj). The beacons operated on frequencies 5471.5 kHz, 7871.5 kHz, 10408.5 kHz, 14396.5 kHz, and 20948.5 kHz. Since 2002, there have been no reception reports for these beacons and the relevant ITU web pages have been removed.

Antenna Farm

The dish farm at the Raisting Satellite Earth Station complex and Telehouse, Germany's largest satellite communications facility in Raisting, Bavaria, Germany.

Antenna farm or satellite dish farm or just dish farm are terms used to describe an area dedicated to television or radio telecommunications transmitting or receiving antenna equipment, such as C, Ku or Ka band satellite dish antennas, UHF/VHF/AM/FM transmitter towers or mobile cell towers. The history of the term "antenna farm" is uncertain, but it dates to at least the 1950s.

In telecom circles, any area with more than three antennas could be referred to as an antenna farm. In the case of an AM broadcasting station (mediumwave and longwave, occasionally shortwave), the multiple mast radiators may all be part of an antenna system for a single station, while for VHF and UHF the site may be under joint management. Alternatively, a single tower with many separate antennas is often called a "candelabra tower".

Safety & Security

Commercial antenna farms are managed by radio stations, television stations, satellite teleports or military organizations and are mostly very secure facilities with access limited to broadcast engineers, RF engineers or maintenance technicians. This is not only for the physical security of the location (including preventing equipment/metal theft), but also for safety, as there may be a radiation hazard unless stations are powered-down.

Locations

Where terrain allows, mountaintop sites are very attractive for non-AM broadcast stations and others, because it increases the stations' height above average terrain, allowing them to reach further by avoiding obstructions on the ground, and by increasing the radio horizon. With a clearer line of sight in both cases, more signal can be received. While the same is true of a very tall tower, such towers are expensive, dangerous, and difficult to access the top of, and may collect and drop large amounts of ice in winter, or even collapse in a severe ice storm and/or high winds. Multiple small towers also allow stations to have backup facilities co-located on each other's towers for redundancy.

Antennas on Mount Wilson, covered in ice

Satellite antenna farms are usually located at remote locations, far away from urban development, especially high rise buildings or airplane flight paths, to avoid and minimize disruption to transmission and reception, and so as to not be an eyesore. Although most radio and TV stations are in fierce competition with each other in their broadcast markets, they will often locate their broadcasting antennas very near each other, and in some cases, will even share land or towers with each other, in the interests of space, land availability, and the cost of putting a transmission building on top of a mountain.

In the United States of America

Most metropolitan areas have at least one antenna farm, such as Mount Wilson in greater Los Angeles, Sweat Mountain in metro Atlanta, Farnsworth Peak for the Salt Lake Valley, Riverview in Tampa, Florida, Baltimore's Television Hill and Slide Mountain (Mount Rose ski area) in the Reno/Tahoe area. Some cities instead have combined many stations onto one tower, often through diplexers into just one or two antennas, such as atop the Empire State Building in New York, the landmark Sutro Tower of San Francisco, or the huge Miami Gardens tower serving the Miami and Fort Lauderdale region. Cleveland, Ohio has its antenna farm in the suburb of Parma, Ohio due to Parma's high elevation. In central Oklahoma City most of the city's media outlets transmitter and tower facilities are located between the Kilpatrick Turnpike to the south and Interstate 44 to the north, Broadway Extension to the west and Interstate 35 to the east with Britton Road being the central thoroughfare. In addition, all three network affiliates and one of the 3 major radio groups have their studio facilities located within the Oklahoma City tower farm.

In the Appalachian Mountains of the Eastern United States, Poor Mountain serves most of the FM and TV stations in the Roanoke/Lynchburg market. Holston Mountain in upper East Tennessee is home to most of the FM and TV stations in the Tri-Cities (Bristol, Virginia-Kingsport, Tennessee-Johnson City, Tennessee) DMA. Other examples are Signal Mountain near Chattanooga, Tennessee, Sharp's Ridge in Knoxville, Tennessee, and Paris Mountain in Greenville, South Carolina.

Other examples of co-located towers on mountain peaks in the United States are on Red Mountain in Birmingham, Alabama; Mount Wilson near Los Angeles; the Sutro Tower in Clarendon Heights near Mount Sutro in San Francisco; Lookout Mountain, Colorado near Denver; Cedar Hill between Dallas and Fort Worth; South Mountain Park near Phoenix; Nelson Peak near Salt Lake City; Sandia Crest near Albuquerque, New Mexico.

Probably the most famous broadcast antenna farm of all is the World Trade Center Tower One, on which many of the New York City television and several FM stations had their antennas. All were lost when Twin Towers One and Two collapsed after the September 11 attacks in 2001. Most of those stations now broadcast from their previous home, 200 feet lower, on the Empire State Building.

Objections

Antennas on Sweat Mountain.

Antenna farms have often been the source of complaints from local neighborhoods, particularly when a new tower is added. This has been increasingly so for TV stations, which have been pursuing with alacrity the construction of new digital television antennas. Because many of these towers are already full, or were built well before there was the expectation of DTV, many stations have been forced to go through the even greater expense of constructing a new tower.

One such situation was in Colorado, in the late 1990s and early to mid-2000s. Many of the Denver metropolitan area TV stations already transmitted analog TV from Lookout Mountain, but needed the extra space for more antennas. Additionally, since many people live on Lookout Mountain, there was also the concern about safety, not only from falling ice or even the slight risk of a tower collapse, but also ongoing from the additional RF that it would create. Residents and the city of Golden filed legal objections, including challenging the authority of the Federal Communications Commission (FCC) to override denial of zoning permits by local government (in this case, the Jefferson county commission). The city of Golden also sought to condemn the site, even though it was outside city limits. It was decided that scenic Eldorado Mountain near Boulder might be a better site, but there were also objections that it would ruin the view of that mountain from the valley. Despite other alternatives, the new "supertower" was forced on Lookout Mountain by the U.S. Congress, allowing the existing towers to be removed in 2009 after analog shutdown. The site began operating in spring 2008.

Antenna Farm Requirements

A radome (left) among multiple Cassegrain satellite antennas located at the RaistingSatellite Earth Station complex.

- Clear line of sight, especially for microwave dishes

- Free of radio interference, such as marine radar

- Higher ground elevation, to maximize coverage

- Away from high-rise structures and other obstructions

- Must be at least 20 miles away from airports.

Isotropic Radiator

An isotropic radiator is a theoretical point source of electromagnetic or sound waves which radiates the same intensity of radiation in all directions. It has no preferred direction of radiation. It radiates uniformly in all directions over a sphere centred on the source. Isotropic radiators are used as reference radiators with which other sources are compared.

Whether a radiator is isotropic is independent of whether it obeys Lambert's law. As radiators, a spherical black body is both, a flat black body is Lambertian but not isotropic, a flat chrome sheet is neither, and by symmetry the Sun is isotropic, but not Lambertian on account of limb darkening.

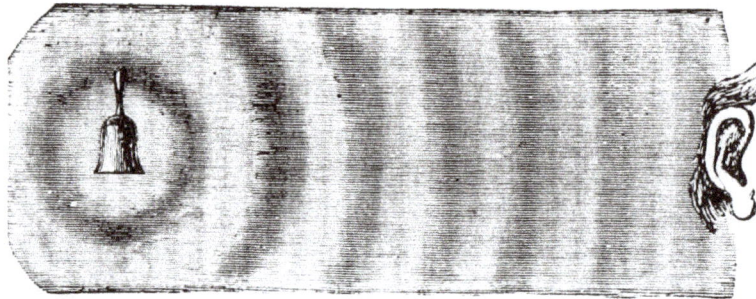

A depiction of an isotropic radiator of sound, published in Popular Science Monthly in 1878. Note how the rings are even and of the same width all the way around each circle, though they fade as they move away from the source.

Physics

In physics, an isotropic radiator is a point radiation or sound source. At a distance, the sun is an isotropic radiator of electromagnetic radiation. The Big Bang is another example of an isotropic radiator - the Cosmic Microwave Background.

Antenna Theory

In antenna theory, an isotropic antenna is a hypothetical antenna radiating the same intensity of radio waves in all directions. It thus is said to have a directivity of 0 dBi (dB relative to isotropic) in all directions.

In reality, a *coherent* isotropic radiator of linear polarization can be shown to be impossible. Its radiation field could not be consistent with the Helmholtz wave equation (derived from Maxwell's equations) in all directions simultaneously. Consider a large sphere surrounding the hypothetical point source, so that at that radius the wave over a reasonable area is essentially planar. The electric (and magnetic) field of a plane wave in free space is always perpendicular to the direction of propagation of the wave. So the electric field would have to be tangent to the surface of the sphere everywhere, and continuous along that surface. However the hairy ball theorem shows that a continuous vector field tangent to the surface of a sphere must fall to zero at one or more points on the sphere, which is inconsistent with the assumption of an isotropic radiator with linear polarization.

Incoherent isotropic radiators are possible and do not violate Maxwell's equations. Acoustic isotropic radiators are possible because sound waves in a gas or liquid are longitudinal waves and not transverse waves.

Even though an isotropic antenna cannot exist in practice, it is used as a base of comparison to calculate the directivity of actual antennas. Antenna gain G, which is equal to the antenna's directivity multiplied by the antenna efficiency, is defined as the ratio of the intensity I (power per unit area) of the radio power received at a given distance from the antenna (in the direction of maximum radiation) to the intensity I_{iso} received from an isotropic antenna at the same distance. This is called *isotropic gain*:

$$G = \frac{I}{I_{iso}}$$

Thus the gain of any perfectly efficient antenna averaged over all directions is unity.

Isotropic Receiver

In EMF measurements applications, an isotropic receiver (also called isotropic antenna) is a field measurement instrument which allows to obtain the total field independently of the tri-axial orthogonal arrangement chosen for orientation of the device itself.

In practice a quasi-ideal isotropic receiver is obtained with three orthogonal sensing devices with a radiation diagram of the omnidirectional type $\sin(\theta)$, like that of short dipole and small loop antennas.

The parameter used to define accuracy in the measurements is called isotropic deviation.

Derivation of Isotropic Antenna Aperture

Imagine two cavities in thermal equilibrium. A lossless antenna in one cavity is connected to a matched resistor inside the second cavity. Using the Rayleigh-Jeans approximation

$$B_v = \frac{2kT}{\lambda^2},$$

the power received by the antenna over a narrow frequency band is

$$P_v = A_e S_{matched} = A_e \frac{S}{2} = \frac{1}{2}\left(\int_{4\pi} A_e B_v d\Omega\right)dv.$$

Since the cavities are in thermal equilibrium, this equals the Nyquist spectral power of the resistor

$$P_v = kTdv$$

in that same frequency band, thus

$$\frac{1}{2}\left(\int_{4\pi} A_e B_v d\Omega\right)dv = kTdv.$$

$$\frac{1}{2}\left(\int_{4\pi} A_e \frac{2kT}{\lambda^2} d\Omega\right)dv = kTdv.$$

Being isotropic, A_e is constant in every direction, thus

$$A_e \frac{kT}{\lambda^2} \int_{4\pi} d\Omega = kT.$$

Optics

$$A_e = \frac{\lambda^2}{4\pi}.$$

In optics, an isotropic radiator is a point source of light. The sun approximates an isotropic radiator of light. Certain munitions such as flares and chaff have isotropic radiator properties.

Sound

An isotropic radiator is a theoretical perfect speaker exhibiting equal sound volume in all directions.

References

- S.M. Lindenmeier, L.M. Reiter, D.E. Barie and J.F. Hopf. "Antenna Diversity for Improving the BER in Mobile Digital Radio Reception Especially in Areas with Dense Foliage." International ITG Conference on Antennas, ISBN 978-3-00-021643-5.

- Bakshi, K.A.; A.V.Bakshi, U.A.Bakshi (2009). Antennas And Wave Propagation. Technical Publications. p. 1.17. ISBN 81-8431-278-4.

- Pat Hawker, G3VA (2008). "The DK0WCY/DRA5 Propagation Beacons". Technical Topics Scrapbook - All 50 years. Potters Bar, UK: Radio Society of Great Britain. p. 98. ISBN 978-1-905086-39-9.

- "IARU Region 2 Band Plan" (PDF). International Amateur Radio Union Region 2. September 27, 2013. Retrieved October 27, 2015.

5

Key Aspects of Propagation

When radio waves travel from one point to another, they exhibit characteristics determined by that location and such behavior is called propagation. Radio propagation can be affected by numerous influences determined by its path from A point to B point. This chapter explicates the key features of propagation by further explaining theories like shortwave radio, surface wave, two-way radio and skip zone.

Shortwave Radio

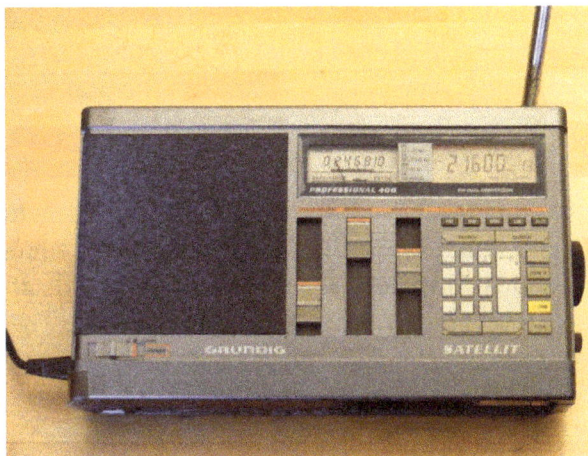

A solid-state, digital shortwave receiver

Shortwave radio is radio transmission using shortwave frequencies, generally 1.6–30 MHz (187.4–10.0 m), just above the medium wave AM broadcast band.

Shortwave radio has an ability to enter a state of *skywave* or *skip* propagation, in which the radio waves are reflected or refracted back to Earth from the ionosphere, as opposed to being absorbed. This is a particularly important characteristic: it allows the transmitted signal to be reflected around the curve of the Earth, thus shortwave is not line-of-sight. This allows for very long distance communications. Shortwave radio is used for broadcasting of voice and music to shortwave listeners over very large areas; sometimes entire continents or beyond. Additionally, it is used for two-way international communication by amateur radio enthusiasts for hobby, educational and emergency purposes.

Frequency Classifications

The widest popular definition of the shortwave frequency interval is the ITU Region 1 (EU+Africa+Russia...) definition, and is the span 1.6–30 MHz, just above the medium wave band, which ends approximately at 1.6 MHz.

There are also other definitions of the shortwave frequency interval:

- 1.71 to 30 MHz in ITU Region 2 (North and South America...)

- 1.8 (160 meter radio amateur band start) to 30 MHz

- 2.3 (120 meter band start) to 30 MHz

- 2.3 (120 meter band start) to 26.1 MHz (11 meter band end)

- In Germany and perhaps Austria the ITU Region 1 shortwave frequency interval can be subdivided in:

  - *de:Grenzwelle* ("border waves"): 1.605-3.8 MHz and *de:Kurzwelle* (shortwaves) 3.8-30 MHz

  - *Grenzwelle*: 1.605-4 MHz and *Kurzwelle* (shortwaves) 4-30 MHz

- In Germany these shortwave frequency intervals have also been seen used:

  - 3-30 MHz – e.g. some accept that "high frequency" is the same as "short wave". In reality, the definition of the "shortwave" frequency band is a mess, and therefore the "shortwave frequencies" can not be exactly equal "high frequencies".

  - the above other definitions

Shortwave radio received its name because the wavelengths in this band are shorter than 200 m (1500 kHz) which marked the original upper limit of the medium frequency band first used for radio communications. The broadcast medium wave band now extends above the 200 m/1500 kHz limit, and the amateur radio 1.8 MHz – 2.0 MHz band (known as the "top band") is the lowest-frequency band considered to be 'shortwave'.

History

Development

Early radio telegraphy had used long wave transmissions. The drawbacks to this system included a very limited spectrum available for long distance communication, and the very expensive transmitters, receivers and gigantic antennas that were required. It was also difficult to beam the radio wave directionally with long wave, resulting in a major loss of power over long distances. Prior to the 1920s, the shortwave frequencies above 2 MHz were regarded as useless for long distance communication and were designated in many countries for amateur use.

Guglielmo Marconi, pioneer of radio, commissioned his assistant Charles Samuel Franklin to carry out a large scale study into the transmission characteristics of short wavelength waves and to determine their suitability for long distance transmissions. Franklin rigged up a large antenna at Poldhu Wireless Station, Cornwall, running on 25 kW of power. In June and July 1923, wireless transmissions were completed during nights on 97 meters from Poldhu to Marconi's yacht *Elettra* in the Cape Verde Islands.

In September 1924, Marconi transmitted daytime and nighttime on 32 meters from Poldhu to his yacht in Beirut. Franklin went on to refine the directional transmission, by inventing the curtain array aerial system. In July 1924, Marconi entered into contracts with the British General Post Office (GPO) to install high speed shortwave telegraphy circuits from London to Australia, India, South Africa and Canada as the main element of the Imperial Wireless Chain. The UK-to-Canada shortwave "Beam Wireless Service" went into commercial operation on 25 October 1926. Beam Wireless Services from the UK to Australia, South Africa and India went into service in 1927.

Shortwave communications began to grow rapidly in the 1920s, similar to the internet in the late 20th century. By 1928, more than half of long distance communications had moved from transoceanic cables and longwave wireless services to shortwave and the overall volume of transoceanic shortwave communications had vastly increased. Shortwave also ended the need for multimillion-dollar investments in new transoceanic telegraph cables and massive longwave wireless stations, although some existing transoceanic telegraph cables and commercial longwave communications stations remained in use until the 1960s.

The cable companies began to lose large sums of money in 1927, and a serious financial crisis threatened the viability of cable companies that were vital to strategic British interests. The British government convened the Imperial Wireless and Cable Conference in 1928 "to examine the situation that had arisen as a result of the competition of Beam Wireless with the Cable Services". It recommended and received Government approval for all overseas cable and wireless resources of the Empire to be merged into one system controlled by a newly formed company in 1929, Imperial and International Communications Ltd. The name of the company was changed to Cable and Wireless Ltd. in 1934.

Amateur Use of Shortwave Propagation

Amateur radio operators also discovered that long-distance communication was possible on shortwave bands. Early long-distance services used surface wave propagation at very low frequencies, which are attenuated along the path. Longer distances and higher frequencies using this method meant more signal attenuation. This, and the difficulties of generating and detecting higher frequencies, made discovery of shortwave propagation difficult for commercial services.

Radio amateurs may have conducted the first successful transatlantic tests in December 1921, operating in the 200 meter mediumwave band (1500 kHz)—the shortest wavelength then available to amateurs. In 1922 hundreds of North American amateurs were heard in Europe at 200 meters and at least 20 North American amateurs heard amateur signals from Europe. The first two-way communications between North American and Hawaiian amateurs began in 1922 at 200 meters. Although operation on wavelengths shorter than 200 meters was technically illegal (but tolerated as the authorities mistakenly believed at first that such frequencies were useless for commercial or military use), amateurs began to experiment with those wavelengths using newly available vacuum tubes shortly after World War I.

Extreme interference at the upper edge of the 150-200 meter band—the official wavelengths allocated to amateurs by the Second National Radio Conference in 1923—forced amateurs to shift to shorter and shorter wavelengths; however, amateurs were limited by regulation to wavelengths longer than 150 meters (2 MHz). A few fortunate amateurs who obtained special permission for experimental communications below 150 meters completed hundreds of long distance two way contacts on 100 meters (3 MHz) in 1923 including the first transatlantic two way contacts.

By 1924 many additional specially licensed amateurs were routinely making transoceanic contacts at distances of 6,000 miles (~9,600 km) and more. On 21 September several amateurs in California completed two way contacts with an amateur in New Zealand. On 19 October amateurs in New Zealand and England completed a 90-minute two-way contact nearly halfway around the world. On October 10, the Third National Radio Conference made three shortwave bands available to U.S. amateurs at 80 meters (3.75 MHz), 40 meters (7 MHz) and 20 meters (14 MHz). These were allocated worldwide, while the 10-meter band (28 MHz) was created by the Washington International Radiotelegraph Conference on 25 November 1927. The 15-meter band (21 MHz) was opened to amateurs in the United States on 1 May 1952.

Propagation Characteristics

Shortwave radio frequency energy is capable of reaching any location on the Earth as it is influenced by ionospheric reflection back to the earth by the ionosphere, (a phenomenon known as "skywave propagation"). A typical phenomenon of shortwave propagation is the occurrence of a skip zone where reception fails. With a fixed working frequency, large changes in ionospheric conditions may create skip zones at night.

As a result of the multi-layer structure of the ionosphere, propagation often simultaneously occurs on different paths, scattered by the E or F region and with different numbers of hops, a phenomenon that may be disturbed for certain techniques. Particularly for lower frequencies of the shortwave band, absorption of radio frequency energy in the lowest ionospheric layer, the D layer, may impose a serious limit. This is due to collisions of electrons with neutral molecules, absorbing some of a radio frequency's energy and converting it to heat. Predictions of skywave propagation depend on:

- The distance from the transmitter to the target receiver.

- Time of day. During the day, frequencies higher than approximately 12 MHz can travel longer distances than lower ones. At night, this property is reversed.

- With lower frequencies the dependence on the time of the day is mainly due to the lowest ionospheric layer, the D Layer, forming only during the day when photons from the sun break up atoms into ions and free electrons.

- Season. During the winter months of the Northern or Southern hemispheres, the AM/MW broadcast band tends to be more favorable because of longer hours of darkness.

- Solar flares produce a large increase in D region ionization so high, sometimes for periods of several minutes, all skywave propagation is nonexistent.

Types of Modulation

Several different types of modulation are used to impress information on a short-wave transmission.

Amplitude modulation is the simplest type and the most commonly used for shortwave broadcasting. The instantaneous amplitude of the carrier is controlled by the amplitude of the signal (speech, or music, for example). At the receiver, a simple detector recovers the desired modulation signal from the carrier.

Single sideband transmission is a form of amplitude modulation but in effect filters the result of modulation. An amplitude-modulated signal has frequency components both above and below the carrier frequency. If one set of these components is eliminated as well as the residual carrier, only the remaining set is transmitted. This saves power in the transmission, as roughly 2/3 of the energy sent by an AM signal is unnecessary to recover the information contained on it. It also saves "bandwidth", allowing about one-half the carrier frequency spacing to be used. The drawback is that the receiver is more complicated, since it must re-recreate the carrier to recover the signal. Small errors in the detector process can greatly affect the pitch of the received signal, so single side band is not usual for music or general broadcast. Single side band is used for long-range voice communications by ships and aircraft, Citizen's Band, and amateur radio operators. LSB (lower sideband) is generally used below 9 MHz and USB (upper sideband) above 9 MHz.

Vestigal sideband transmits the carrier and one complete side-band, but filters out the redundant side-band. It is a compromise between AM and SSB, allowing simple receivers to be used but requiring almost as much transmitter power as AM. One advantage is that only half the bandwidth of an AM signal is used. It can be heard in the transmission of certain radio time signal stations.

Continuous wave (CW) is on-and-off keying of a carrier, used only for Morse code communications.

Narrow-band frequency modulation (NBFM) is mainly used in the higher HF frequencies (typically above 20 MHz). Because of the larger bandwidth required, NBFM is much more commonly used for VHF communication. Regulations limit the bandwidth of a signal transmitted in the HF bands, and the advantages of frequency modulation are greatest if the FM signal is allowed to have a wider bandwidth. NBFM is limited to short-range SW transmissions due to the multiphasic distortions created by the ionosphere.

Digital Radio Mondiale (DRM) is a digital modulation for use on bands below 30 MHz.

Radioteletype, fax, digital, slow-scan television and other systems use forms of frequency-shift keying or audio subcarriers on a shortwave carrier. These generally require special equipment to decode, such as software on a computer equipped with a sound card.

Uses

Some major uses of the shortwave radio band are:

- International broadcasting primarily by government-sponsored propaganda, international news (for example, the BBC World Service) or cultural stations to foreign audiences: the most common use of all.

- Domestic broadcasting: to widely dispersed populations with few longwave, mediumwave and FM stations serving them; or for specialty political, religious and alternative media networks; or of individual commercial and non-commercial paid broadcasts.

- Oceanic air traffic control uses the HF/shortwave band for long distance communication to aircraft over the oceans and poles, which are far beyond the range of traditional VHF frequencies. Modern systems also include satellite communications, such as ADS-C/CPDLC

- "Utility" stations transmitting messages not intended for the general public, such as aircraft flying between continents, encrypted diplomatic messages, weather reporting, or ships at sea.

- Clandestine stations. These are stations that broadcast on behalf of various political movements, including rebel or insurrectionist forces, and are normally unauthorised by the government-in-charge of the country in question. Clandestine broadcasts may emanate from transmitters located in rebel-controlled territory or from outside the country entirely, using another country's transmission facilities. Clandestine stations were used during World War II to transmit news from the Allied point of view into Axis-controlled areas. Although the Nazis confiscated many radios and executed their owners, many people continued to listen.

- Numbers Stations These stations regularly appear and disappear all over the shortwave radio band but are unlicensed and untraceable. It is believed that Numbers Stations are operated by government agencies, and are used to communicate with clandestine operatives working within foreign countries. However, no definitive proof of such use has emerged. Because the vast majority of these broadcasts contain nothing but the recitation of blocks of numbers, in various languages, with occasional bursts of music, they have become known colloquially as "Number Stations". Perhaps the most noted Number Station is the "Lincolnshire Poacher", named after the 18th century English folk song, which is transmitted just before the sequences of numbers.

- Amateur radio operators at the 80/75, 60, 40, 30, 20, 17, 15, 12, and 10-meter bands.

- Time signal and radio clock stations: In North America, WWV radio and WWVH radio transmit at these frequencies: 2500 kHz, 5000 kHz, 10000 kHz, and 15000 kHz; and WWV also transmits on 20000 kHz. The CHU radio station in Canada transmits on the following frequencies: 3330 kHz, 7850 kHz, and 14670 kHz. Other similar radio clock stations transmit on various shortwave and longwave frequencies around the world. The shortwave transmissions are primarily intended for human reception, while the longwave stations are generally used for automatic synchronization of watches and clocks.

- Over-the-horizon radar: From 1976 to 1989, the Soviet Union's Russian Woodpecker over-the-horizon radar system blotted out numerous shortwave broadcasts daily.

The term DXing, in the context of listening to radio signals of any user of the shortwave band, is the activity of monitoring distant stations. In the context of amateur radio operators, the term "DXing" refers to the two-way communications with a distant station, using shortwave radio frequencies.

The Asia-Pacific Telecommunity estimates that there are approximately 600,000,000 shortwave broadcast-radio receivers in use in 2002. WWCR claims that there are 1.5 billion shortwave receivers worldwide.

Shortwave Broadcasting

Frequency Allocations

The World Radiocommunication Conference (WRC), organized under the auspices of the International Telecommunication Union, allocates bands for various services in conferences every few years. The last WRC took place in 2007.

At WRC-97 in 1997, the following bands were allocated for international broadcasting. AM shortwave broadcasting channels are allocated with a 5 kHz separation for traditional analog audio broadcasting.

Metre Band	Frequency Range	Remarks
120 m	2.3–2.495 MHz	tropical band
90 m	3.2–3.4 MHz	tropical band
75 m	3.9–4 MHz	shared with the North American amateur radio 80m band
60 m	4.75–5.06 MHz	tropical band
49 m	5.9–6.2 MHz	
41 m	7.2–7.6 MHz	shared with the amateur radio 40m band
31 m	9.4–9.9 MHz	currently the most heavily used band
25 m	11.6-12.2 MHz	
22 m	13.57-13.87 MHz	
19 m	15.1-15.8 MHz	
16 m	17.48-17.9 MHz	
15 m	18.9-19.02 MHz	almost unused, could become a DRM band
13 m	21.45-21.85 MHz	
11 m	25.6-26.1 MHz	may be used for local DRM broadcasting

Although countries generally follow the table above, there may be small differences between countries or regions. For example, in the official bandplan of the Netherlands, the 49 m band starts at 5.95 MHz, the 41 m band ends at 7.45 MHz, the 11 m band starts at 25.67 MHz, and the 120, 90 and 60 m bands are absent altogether. Additionally, international broadcasters sometimes operate outside the normal WRC-allocated bands or use off-channel frequencies. This is done for practical reasons, or to attract attention in crowded bands (60m, 49m, 40m, 41m, 31m, 25m).

The new digital audio broadcasting format for shortwave DRM operates 10 kHz or 20 kHz channels. There are some ongoing discussions with respect to specific band allocation for DRM, as it mainly transmitted in 10 kHz format.

The power used by shortwave transmitters ranges from less than one watt for some experimental and amateur radio transmissions to 500 kilowatts and higher for intercontinental broadcasters and over-the-horizon radar. Shortwave transmitting centers often use specialized antenna designs (like the ALLISS antenna technology) to concentrate radio energy at the target area.

Advantages

Shortwave does possess a number of advantages over newer technologies, including the following:

- Difficulty of censoring programming by authorities in restrictive countries: unlike their relative ease in monitoring the Internet, government authorities face technical difficulties monitoring which stations (sites) are being listened to (accessed). For example, during the

Russian coup against President Mikhail Gorbachev, when his access to communications was limited, Gorbachev was able to stay informed by means of the BBC World Service on shortwave.

- Low-cost shortwave radios are widely available in all but the most repressive countries in the world. Simple shortwave regenerative receivers can be easily built with a few parts.

- In many countries (particularly in most developing nations and in the Eastern bloc during the Cold War era) ownership of shortwave receivers has been and continues to be widespread (in many of these countries some domestic stations also used short-wave).

- Many newer shortwave receivers are portable and can be battery-operated, making them useful in difficult circumstances. Newer technology includes hand-cranked radios which provide power without batteries.

- Shortwave radios can be used in situations where Internet or satellite communications service is temporarily or long-term unavailable (or unaffordable).

- Shortwave radio travels much farther than broadcast FM (88-108 MHz). Shortwave broadcasts can be easily transmitted over a distance of several thousands of kilometers, including from one continent to another.

- Particularly in tropical regions, SW is somewhat less prone to interference from thunderstorms than medium wave radio, and is able to cover a large geographic area with relatively low power (and hence cost). Therefore, in many of these countries it is widely used for domestic broadcasting.

- Very little infrastructure is required for long-distance two-way communications using shortwave radio. All one needs is a pair of transceivers, each with an antenna, and a source of energy (such as a battery, a portable generator, or the electrical grid). This makes shortwave radio one of the most robust means of communications, which can be disrupted only by interference or bad ionospheric conditions. Modern digital transmission modes such as MFSK and Olivia are even more robust, allowing successful reception of signals well below the noise floor of a conventional receiver.

Disadvantages

Shortwave radio's benefits are sometimes regarded as being outweighed by its drawbacks, including:

- In most Western countries, shortwave radio ownership is usually limited to true enthusiasts, since most new standard radios do not receive the shortwave band. Therefore, Western audiences are limited.

- In the developed world, shortwave reception is very difficult in urban areas because of excessive noise from switched-mode power adapters, fluorescent or LED light sources, internet modems and routers, computers and many, many other sources of radio interference.

Shortwave Listening

A pennant sent to overseas listeners by Radio Budapest in the late 1980s

Many hobbyists listen to shortwave broadcasters without operating their own transmitters. In some cases, the goal is to hear as many stations from as many countries as possible *(DXing)*; others listen to specialized shortwave utility, or "ute", transmissions such as maritime, naval, aviation, or military signals. Others focus on intelligence signals from numbers stations, stations which transmit strange broadcast usually for intelligence operations, or the two way communications by amateur radio operators. Some short wave listeners behave analogously to "lurkers" on the Internet, in that they listen only and never make any attempt to send out their own signals. Other listeners participate in clubs, or actively send and receive QSL cards, or become involved with amateur radio and start transmitting on their own.

Many listeners tune the shortwave bands for the programmes of stations broadcasting to a general audience (such as Radio Taiwan International, Voice of Russia, China Radio International, Radio Canada International, Voice of America, Radio France Internationale, BBC World Service, Radio Australia, Radio Netherlands, Voice of Korea, Radio Free Sarawak etc.). Today, through the evolution of the Internet, the hobbyist can listen to shortwave signals via remotely controlled shortwave receivers around the world, even without owning a shortwave radio. Many international broadcasters (such as Radio Canada International , the BBC and Radio Australia) offer live streaming audio on their websites. Shortwave listeners, or SWLs, can obtain QSL cards from broadcasters, utility stations or amateur radio operators as trophies of the hobby. Some stations even give out special certificates, pennants, stickers and other tokens and promotional materials to shortwave listeners.

Amateur Radio

The practice of operating a shortwave radio transmitter for non-commercial two-way communications is known as amateur radio. Licenses are granted by authorized government agencies.

Amateur radio operators have made many technical advancements in the field of radio, and make themselves available to transmit emergency communications when normal communications channels fail. Some amateurs practice operating *off the power grid* so as to be prepared for power loss. Many amateur radio operators started out as Shortwave Listeners (SWLs) and actively encourage SWLs to become amateur radio operators.

Utility Stations

Utility stations are stations that do not intentionally broadcast to the general public (although their signals can be received by anybody with appropriate equipment). There are shortwave bands allocated to the use of merchant shipping, marine weather, and ship-to-shore stations; for aviation weather and air-to-ground communications; for military communications; for long-distance governmental purposes, and for other non-broadcast communications. Many radio hobbyists specialize in listening to "ute" broadcasts, which often originate from geographic locations without known shortwave broadcasters.

Unusual Signals

The short wave bands are also used by unlicensed individuals who may want mostly short-range "party line" like communications. Two examples are the use of HF for communication between fishing boats in many areas of the world, and the unlicensed use of the 11-meter band, which is effectively permitted in some areas of the world. Unlicensed operators, called "pirates", can cause signal interference to licensed stations. Many third-world countries have shops selling HF transmitter radios to any customer without regard to license or operator knowledge. As of 2012, there were virtually no national or international efforts to control such pirate operations.

The short wave bands are also used for various experiments, some continuing for years. In 2011, signals traceable to China regularly sent powerful HF transmissions scanning wide ranges of HF frequencies, perhaps to determine the maximum usable frequency (MUF) or other variables.

Numbers stations are broadcasts on shortwave radio that are coded into groups of numbers. Their content is generally encrypted and their purpose remains a mystery.

Shortwave Broadcasts and Music

Some musicians have been attracted to the unique aural characteristics of shortwave radio which—due to the nature of amplitude modulation, varying propagation conditions, and the presence of interference—generally has lower fidelity than local broadcasts (particularly via FM stations). Shortwave transmissions often have bursts of distortion, and "hollow" sounding loss of clarity at certain aural frequencies, altering the harmonics of natural sound and creating at times a strange "spacey" quality due to echoes and phase distortion. Evocations of shortwave reception distortions have been incorporated into rock and classical compositions, by means of delays or feedback loops, equalizers, or even playing shortwave radios as live instruments. Snippets of broadcasts have been mixed into electronic sound collages and live musical instruments, by means of analogue tape loops or digital samples. Sometimes the sounds of instruments and existing musical recordings are altered by remixing or equalizing, with various distortions added, to replicate the garbled effects of shortwave radio reception.

The first attempts by serious composers to incorporate radio effects into music may be those of the Russian physicist and musician Léon Theremin, who perfected a form of radio oscillator as a musical instrument in 1928 (regenerative circuits in radios of the time were prone to breaking into oscillation, adding various tonal harmonics to music and speech); and in the same year, the development of a French instrument called the Ondes Martenot by its inventor Maurice Martenot, a French cellist and former wireless telegrapher. A notable chamber piece by Mexican composer Silvestre Revueltas—*Ocho x radio*, 1933—features a complex texture of pseudo-mariachi musics, overlapping and cross-fading as if heard from distant stations: quite similar to shortwave radio signal propagation disturbance. John Cage used actual radios (of unspecified wavelength) live on several occasions, starting in 1942 with *Credo in Us*, while Karlheinz Stockhausen used shortwave radio and effects in works including *Hymnen* (1966–67), *Kurzwellen* (1968)—adapted for the Beethoven Bicentennial in *Opus 1970* with filtered and distorted snippets of Beethoven pieces—*Spiral* (1968), *Pole, Expo* (both 1969–70), and *Michaelion* (1997).

Holger Czukay, a student of Stockhausen, was one of the first to use shortwave in a rock music context. In 1975, German electronic music band Kraftwerk recorded a full length concept album around simulated radiowave and shortwave sounds, entitled *Radio-Activity*. Among others, The The whose Radio Cineola monthly broadcasts draw heavily on shortwave radio sound, The B-52s, Shearwater, Tom Robinson, Peter Gabriel, Pukka Orchestra, AMM, John Duncan, Orchestral Manoeuvres in the Dark (on their *Dazzle Ships* album), Pat Metheny, Aphex Twin, Boards of Canada, PressureWorks, Rush, Able Tasmans, Team Sleep, Underworld, Meat Beat Manifesto, Tim Hecker, Jonny Greenwood of Radiohead, Roger Waters (on *Radio KAOS* album), Wilco, code 000 and Samuel Trim have also used or been inspired by shortwave broadcasts.

Shortwave's Future

The development of direct broadcasts from satellites has reduced the demand for shortwave receiver hardware, but there are still a great number of shortwave broadcasters. A new digital radio technology, Digital Radio Mondiale (DRM), is expected to improve the quality of shortwave audio from very poor to standards comparable to the FM broadcast band. The future of shortwave radio is threatened by the rise of power line communication (PLC), also known as Broadband over Power Lines (BPL), which uses a data stream transmitted over unshielded power lines. As the BPL frequencies used overlap with shortwave bands, severe distortions can make listening to analog shortwave radio signals near power lines difficult or impossible. However, because shortwave is a cheap and effective way to receive communications in countries with poor infrastructure, it will be around for years to come.

Shortwave use by hobbyists and licensed amateur ham radio operators continues, and after declining interest for a few years due to competing interests in computers and other communication devices, a new resurgence of interest has occurred as evidenced by the increase of new amateur operator licenses issued worldwide. Some hobbyists have combined amateur radio HF with computers for experimental and established data modes that can communicate very close to under the noise floor of receivers - e.g. WSJT, WSPR.

Surface Wave

Diving grebe creates surface waves

In physics, a surface wave is a mechanical wave that propagates along the interface between differing media, usually as a gravity wave between two fluids with different densities. A surface wave can also be an elastic (or a seismic) wave, such as with a *Rayleigh* or *Love* wave. It can also be an electromagnetic wave guided by a refractive index gradient. In radio transmission, a ground wave is a surface wave that propagates close to the surface of the Earth.

Mechanical Waves

In seismology, several types of surface waves are encountered. Surface waves, in this mechanical sense, are commonly known as either *Love waves* (L waves) or *Rayleigh waves*. A seismic wave is a wave that *travels through the Earth, often as the result of an earthquake or explosion*. Love waves have transverse motion (movement is perpendicular to the direction of travel, like light waves), whereas Rayleigh waves have both longitudinal (movement parallel to the direction of travel, like sound waves) and transverse motion. Seismic waves are studied by seismologists and measured by a seismograph or seismometer. Surface waves span a wide frequency range, and the period of waves that are most damaging is usually 10 seconds or longer. Surface waves can travel around the globe many times from the largest earthquakes. Surface waves are caused when P waves and S waves come to the surface.

The term "surface wave" can describe waves over an ocean, even when they are approximated by Airy functions and are more properly called creeping waves. Examples are the waves at the surface of water and air (ocean surface waves), or ripples in the sand at the interface with water or air. Another example is internal waves, which can be transmitted along the interface of two water masses of different densities.

In theory of Hearing physiology, the Traveling Wave (TW) of Von Bekesy, resulted from an acoustic surface wave of the basilar membrane into the cochlear duct. His theory pretended to explain every features of the auditory sensation owing to these passive mechanical phenomena. But Jozef Zwislocki and later David Kemp, showed that that was irrealistic and that an active feedback was necessary.

Electromagnetic Waves

Ground waves refer to the propagation of radio waves parallel to and adjacent to the surface of the Earth, following the curvature of the Earth. These *surface waves* are also known loosely as the Norton surface wave, the Zenneck surface wave, Sommerfeld waves, and gliding waves.

Radio Propagation

Lower frequencies, below 3 MHz, travel efficiently as ground waves. In ITU nomenclature, this includes (in order): medium frequency (MF), low frequency (LF), very low frequency (VLF), ultra low frequency (ULF), super low frequency (SLF), extremely low frequency (ELF) waves.

Ground propagation works because lower-frequency waves are more strongly diffracted around obstacles due to their long wavelengths, allowing them to follow the Earth's curvature. The Earth has one refractive index and the atmosphere has another, thus constituting an interface that supports the surface wave transmission. Ground waves propagate in vertical polarization, with their magnetic field horizontal and electric field (close to) vertical. With VLF waves, the Ionosphere and earth's surface act as a waveguide.

Conductivity of the surface affects the propagation of ground waves, with more conductive surfaces such as sea water providing better propagation. Increasing the conductivity in a surface results in less dissipation. The refractive indices are subject to spatial and temporal changes. Since the ground is not a perfect electrical conductor, ground waves are attenuated as they follow the earth's surface. The wavefronts initially are vertical, but the ground, acting as a lossy dielectric, causes the wave to tilt forward as it travels. This directs some of the energy into the earth where it is dissipated, so that the signal decreases exponentially.

Most long-distance LF "longwave" radio communication (between 30 kHz and 300 kHz) is a result of groundwave propagation. Mediumwave radio transmissions (frequencies between 300 kHz and 3000 kHz), including AM broadcast band, travel both as groundwaves and, for longer distances at night, as skywaves. Ground losses become lower at lower frequencies, greatly increasing the coverage of AM stations using the lower end of the band. The VLF and LF frequencies are mostly used for military communications, especially with ships and submarines. The lower the frequency the better the waves penetrate sea water. ELF waves (below 3 kHz) have even been used to communicate with deeply submerged submarines.

Surface waves have been used in over-the-horizon radar, which operates mainly at frequencies between 2 and 20 MHz over the sea, which has a sufficiently high conductivity to convey the surface waves to and from a reasonable distance (up to 100 km or more; over-horizon radar also uses skywave propagation at much greater distances). In the development of radio, surface waves were used extensively. Early commercial and professional radio services relied exclusively on long wave, low frequencies and ground-wave propagation. To prevent interference with these services, amateur and experimental transmitters were restricted to the high frequencies (HF), felt to be useless since their ground-wave range was limited. Upon discovery of the other propagation modes possible at medium wave and short wave frequencies, the advantages of HF for commercial and military purposes became apparent. Amateur experimentation was then confined only to authorized frequencies in the range.

Mediumwave and shortwave reflect off the ionosphere at night, which is known as skywave. During daylight hours, the lower D layer of the ionosphere forms and absorbs lower frequency energy. This prevents skywave propagation from being very effective on mediumwave frequencies in daylight hours. At night, when the D layer dissipates, mediumwave transmissions travel better by skywave. Ground waves *do not* include ionospheric and tropospheric waves.

The propagation of sound waves through the ground taking advantage of the earths ability to more efficiently transmit low frequency is known as Audio ground wave (AGW).

Microwave Field Theory

Within microwave field theory, the interface of a dielectric and conductor supports "surface wave transmission". Surface waves have been studied as part of transmission lines and some may be considered as single-wire transmission lines.

Characteristics and utilizations of the electrical surface wave phenomenon include:

- The field components of the wave diminish with distance from the interface.

- Electromagnetic energy is not converted from the surface wave field to another form of energy (except in leaky or lossy surface waves) such that the wave does not transmit power normal to the interface, i.e. it is evanescent along that dimension.

- In optical fiber transmission, evanescent waves are surface waves.

- In coaxial cable in addition to the TEM mode there also exists a transverse-magnetic (TM) mode which propagates as a surface wave in the region around the central conductor. For coax of common impedance this mode is effectively suppressed but in high impedance coax and on a single central conductor without any outer shield, low attenuation and very broadband propagation is supported. Transmission line operation in this mode is called E-Line.

Skywave

In radio communication, skywave or skip refers to the propagation of radio waves reflected or refracted back toward Earth from the ionosphere, an electrically charged layer of the upper atmosphere. Since it is not limited by the curvature of the Earth, skywave propagation can be used to communicate beyond the horizon, at intercontinental distances. It is mostly used in the shortwave frequency bands.

As a result of skywave propagation, a signal from a distant AM broadcasting station, a shortwave station, or—during sporadic E propagation conditions (principally during the summer months in both hemispheres)—a low frequency television station can sometimes be received as clearly as local stations. Most long-distance shortwave (high frequency) radio communication—between 3 and 30 MHz—is a result of skywave propagation. Since the early 1920s amateur radio operators (or "hams"), limited to lower transmitter power than broadcast stations, have taken advantage of skywave for long distance (or "DX") communication.

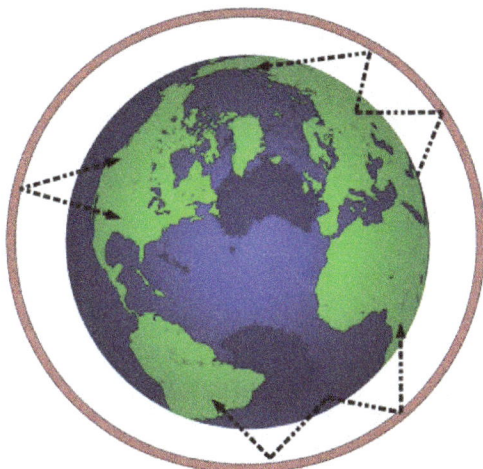

Radio waves (black) reflecting off the ionosphere (red) during skywave propagation.

Skywave propagation is distinct from:

- groundwave propagation, where radio waves travel near Earth's surface without being reflected or refracted by the atmosphere—the dominant propagation mode at lower frequencies,

- line-of-sight propagation, in which radio waves travel in a straight line, the dominant mode at higher frequencies.

Explanation

The ionosphere is a region of the upper atmosphere, from about 80 km to 1000 km in altitude, where neutral air is ionized by solar photons and cosmic rays. When high frequency signals enter the ionosphere obliquely, they are back-scattered from the ionized layer as scatter waves. If the midlayer ionization is strong enough compared to the signal frequency, a scatter wave can exit the bottom of the layer earthwards as if reflected from a mirror. Earth's surface (ground or water) then diffusely reflects the incoming wave back towards the ionosphere. Consequently, like a rock "skipping" across water, the signal may effectively "bounce" or "skip" between the earth and ionosphere two or more times (multihop propagation). Since at shallow incidence losses remain quite small, signals of only a few watts can sometimes be received many thousands of miles away as a result. This is what enables shortwave broadcasts to travel all over the world.

If the ionization is not great enough, the scatter wave is initially deflected downwards, and subsequently upwards (above the layer peak) such that it exits the top of the layer slightly displaced. Sky wave propagation occurs in the waveguide formed by the ground and ionosphere, each serving as reflectors. With a single "hop," path distances up to 3500 km may be reached. Transatlantic connections are mostly obtained with two or three hops.

The layer of ionospheric plasma with equal ionization (the reflective surface) is not fixed, but undulates like the surface of the ocean. Varying reflection efficiency from this changing surface can cause the reflected signal strength to change, causing *"fading"* in shortwave broadcasts.

Depending on the transmitting antenna, signals below approximately 10 MHz during the day and 5 MHz at night, entering the ionosphere at a steep angle (near-vertical incidence) may be back-scattered down to Earth within a short range. Alternatively, signals beamed close to the horizon enter the ionosphere at a shallow angle and return to Earth over medium to long distances.

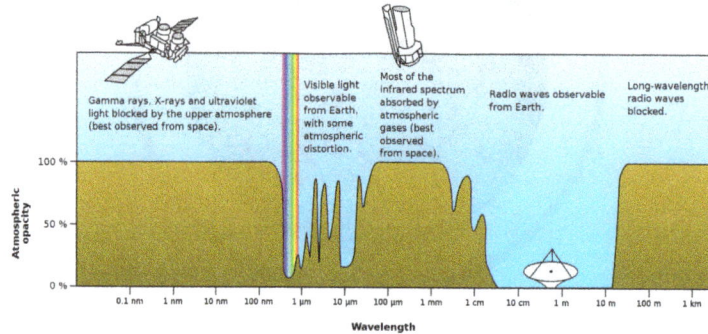

Rough plot of Earth's atmospheric transmittance (or opacity) to various wavelengths of electromagnetic radiation, including radio waves.

Other Considerations

VHF signals with frequencies above about 30 MHz usually penetrate the ionosphere and are not returned to the Earth's surface. E-skip is a notable exception, where VHF signals including FM broadcast and VHF TV signals are frequently reflected to the Earth during late Spring and early Summer. E-skip rarely affects UHF frequencies, except for very rare occurrences below 500 MHz.

Frequencies below approximately 10 MHz (wavelengths longer than 30 meters), including broadcasts in the mediumwave and shortwave bands (and to some extent longwave), propagate most efficiently by skywave at night. Frequencies above 10 MHz (wavelengths shorter than 30 meters) typically propagate most efficiently during the day. Frequencies lower than 3 kHz have a wavelength longer than the distance between the Earth and the ionosphere. The maximum usable frequency for skywave propagation is strongly influenced by sunspot number.

Skywave propagation is usually degraded—sometimes seriously—during geomagnetic storms. Skywave propagation on the sunlit side of the Earth can be entirely disrupted during sudden ionospheric disturbances.

Because the lower-altitude layers (the E-layer in particular) of the ionosphere largely disappear at night, the refractive layer of the ionosphere is much higher above the surface of the Earth at night. This leads to an increase in the "skip" or "hop" distance of the skywave at night.

History

Discovery of Skywave Propagation

Amateur radio operators are credited with the discovery of skywave propagation on the shortwave bands. Early long-distance services used surface wave propagation at very low frequencies, which are attenuated along the path. Longer distances and higher frequencies using this method meant more signal attenuation. This, and the difficulties of generating and detecting higher frequencies, made discovery of shortwave propagation difficult for commercial services.

Radio amateurs conducted the first successful transatlantic tests in December 1921, operating in the 200 meter mediumwave band (1500 kHz)—the shortest wavelength then available to amateurs. In 1922 hundreds of North American amateurs were heard in Europe at 200 meters and at least 30 North American amateurs heard amateur signals from Europe. The first two-way communications between North American and Hawaiian amateurs began in 1922 at 200 meters. Although operation on wavelengths shorter than 200 meters was technically illegal (but tolerated as the authorities mistakenly believed at first that such frequencies were useless for commercial or military use), amateurs began to experiment with those wavelengths using newly available vacuum tubes shortly after World War I.

Extreme interference at the upper edge of the 150-200 meter band—the official wavelengths allocated to amateurs by the Second National Radio Conference in 1923—forced amateurs to shift to shorter and shorter wavelengths; however, amateurs were limited by regulation to wavelengths longer than 150 meters (2 MHz). A few fortunate amateurs who obtained special permission for experimental communications below 150 meters completed hundreds of long distance two way contacts on 100 meters (3 MHz) in 1923 including the first transatlantic two way contacts in November 1923, on 110 meters (2.72 MHz)

By 1924 many additional specially licensed amateurs were routinely making transoceanic contacts at distances of 6000 miles (~9600 km) and more. On 21 September several amateurs in California completed two way contacts with an amateur in New Zealand. On 19 October amateurs in New Zealand and England completed a 90-minute two-way contact nearly halfway around the world. On October 10, the Third National Radio Conference made three shortwave bands available to U.S. amateurs at 80 meters (3.75 MHz), 40 meters (7 MHz) and 20 meters (14 MHz). These were allocated worldwide, while the 10-meter band (28 MHz) was created by the Washington International Radiotelegraph Conference on 25 November 1927. The 15-meter band (21 MHz) was opened to amateurs in the United States on 1 May 1952.

Marconi

In June and July 1923, Guglielmo Marconi's transmissions were completed during nights on 97 meters from Poldhu Wireless Station, Cornwall, to his yacht Ellette in the Cape Verde Islands. In September 1924, Marconi transmitted during daytime and nighttime on 32 meters from Poldhu to his yacht in Beirut. Marconi, in July 1924, entered into contracts with the British General Post Office (GPO) to install high speed shortwave telegraphy circuits from London to Australia, India, South Africa and Canada as the main element of the Imperial Wireless Chain. The UK-to-Canada shortwave "Beam Wireless Service" went into commercial operation on 25 October 1926. Beam Wireless Services from the UK to Australia, South Africa and India went into service in 1927.

Far more spectrum is available for long distance communication in the shortwave bands than in the long wave bands; and shortwave transmitters, receivers and antennas were orders of magnitude less expensive than the multi-hundred kilowatt transmitters and monstrous antennas needed for long wave.

Shortwave communications began to grow rapidly in the 1920s, similar to the internet in the late 20th century. By 1928, more than half of long distance communications had moved from trans-oceanic cables and long wave wireless services to shortwave "skip" transmission and the overall volume of transoceanic shortwave communications had vastly increased. Shortwave also ended the need for multimillion-dollar investments in new transoceanic telegraph cables and massive long wave wireless stations, although some existing transoceanic telegraph cables and commercial long wave communications stations remained in use until the 1960s.

The cable companies began to lose large sums of money in 1927, and a serious financial crisis threatened the viability of cable companies that were vital to strategic British interests. The British government convened the Imperial Wireless and Cable Conference in 1928 "to examine the situation that had arisen as a result of the competition of Beam Wireless with the Cable Services". It recommended and received Government approval for all overseas cable and wireless resources of the Empire to be merged into one system controlled by a newly formed company in 1929, Imperial and International Communications Ltd. The name of the company was changed to Cable and Wireless Ltd. in 1934.

Sporadic E Propagation

Sporadic E or E_s is an unusual form of radio propagation using characteristics of the Earth's ionosphere. Whereas most forms of skywave propagation use the normal and cyclic ionization properties of the ionosphere's F region to refract (or "bend") radio signals back toward the Earth's surface, sporadic E propagation bounces signals off smaller "clouds" of unusually ionized atmospheric gas in the lower E region (located at altitudes of approx. 90 to 160 km). This occasionally allows for long-distance communication at VHF frequencies not usually well-suited to such communication.

Communication distances of 800–2200 km can occur using a single E_s cloud. This variability in distance depends on a number of factors, including cloud height and density. MUF also varies widely, but most commonly falls in the 27–110 MHz range, which includes the FM broadcast band (87.5–108 MHz), Band I VHF television (American channels 2-6, Russian channels 1-3, and European channels 2-4, the latter no longer widely used in Western Europe), CB radio (27 MHz) and the amateur radio 10- and 6-meter bands. Strong events have allowed propagation at frequencies as high as 250 MHz.

As its name suggests, sporadic E is an abnormal event, not the usual condition, but can happen at almost any time; it does, however, display seasonal patterns. Sporadic E activity peaks predictably in the summertime in both hemispheres. In North America, the peak is most noticeable in mid-to-late June, trailing off through July and into August. A much smaller peak is seen around the winter solstice. Activity usually begins in mid-December in the southern hemisphere, with the days immediately after Christmas being the most active period.

On June 12, 2009, sporadic E allowed some television viewers in the eastern United States to see VHF analog TV stations from other states at great distances, in places and on TV channels where local stations had already done their permanent analog shutdown on the final day of the DTV transition in the United States. This was possible because VHF has been mostly avoided by digital TV stations, leaving the analog stations the last ones on the band. It is still possible (as of April, 2010) for many Americans to see Canadian and Mexican analog stations in this manner when sporadic-E occurs, until those countries do their own analog shutdowns over the following few years.

Characteristics

Television and FM signals received via Sporadic E can be extremely strong and range in strength over a short period from just detectable to overloading. Although polarisation shift can occur, single-hop Sporadic E signals tend to remain in the original transmitted polarisation. Long single-hop (900–1,500 miles or 1,400–2,400 kilometres) Sporadic E television signals tend to be more stable and relatively free of multipath images. Shorter-skip (400–800 miles or 640–1,290 kilometres) signals tend to be reflected from more than one part of the Sporadic E layer, resulting in multiple images and ghosting, with phase reversal at times. Picture degradation and signal-strength attenuation increases with each subsequent Sporadic E hop.

Sporadic E usually affects the lower VHF band I (TV channels 2–6) and band II (88–108 MHz FM broadcast band). The typical expected distances are about 600 to 1,400 miles (970 to 2,250 km). However, under exceptional circumstances, a highly ionized Es cloud can propagate band I VHF signals down to approximately 350 miles (560 km). When short-skip Es reception occurs, i.e., under 500 miles (800 km) in band I, there is a greater possibility that the ionized Es cloud will be capable of reflecting a signal at a much higher frequency—i.e., a VHF band 3 channel—since a sharp reflection angle (short skip) favours low frequencies, a shallower reflection angle from the same ionized cloud will favour a higher frequency.

At polar latitudes, Sporadic E can accompany auroras and associated disturbed magnetic conditions and is called Auroral-E.

No conclusive theory has yet been formulated as to the origin of Sporadic E. Attempts to connect the incidence of Sporadic E with the eleven-year Sunspot cycle have provided tentative correlations. There seems to be a positive correlation between sunspot maximum and Es activity in Europe. Conversely, there seems to be a negative correlation between maximum sunspot activity and Es activity in Australasia.

Equatorial E-skip

Equatorial E-skip is a regular daytime occurrence over the equatorial regions and is common in the temperate latitudes in late spring, early summer and, to a lesser degree, in early winter. For receiving stations located within +/− 10 degrees of the geomagnetic equator, equatorial E-skip can be expected on most days throughout the year, peaking around midday local time.

Multipath Propagation

FIn wireless telecommunications, multipath is the propagation phenomenon that results in radio signals reaching the receiving antenna by two or more paths. Causes of multipath include atmospheric ducting, ionospheric reflection and refraction, and reflection from water bodies and terrestrial objects such as mountains and buildings.

Multipath causes multipath interference including constructive and destructive interference, and phase shifting of the signal. Destructive interference causes fading. Where the magnitudes of the signals arriving by the various paths have a distribution known as the Rayleigh distribution, this is known as Rayleigh fading. Where one component (often, but not necessarily, a line of sight component) dominates, a Rician distribution provides a more accurate model, and this is known as Rician fading.

Examples

In facsimile and (analog) television transmission, multipath causes jitter and ghosting, seen as a faded duplicate image to the right of the main image. Ghosts occur when transmissions bounce off a mountain or other large object, while also arriving at the antenna by a shorter, direct route, with the receiver picking up two signals separated by a delay.

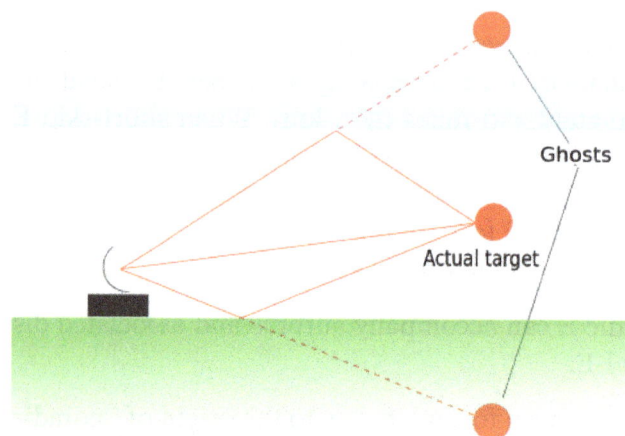

Radar multipath echoes from an actual target cause ghosts to appear.

In radar processing, multipath causes ghost targets to appear, deceiving the radar receiver. These ghosts are particularly bothersome since they move and behave like the normal targets (which they echo), and so the receiver has difficulty in isolating the correct target echo. These problems can be overcome by incorporating a ground map of the radar's surroundings and eliminating all echoes which appear to originate below ground or above a certain height.

In digital radio communications (such as GSM) multipath can cause errors and affect the quality of communications. The errors are due to intersymbol interference (ISI). Equalisers are often used to correct the ISI. Alternatively, techniques such as orthogonal frequency division modulation and rake receivers may be used.

In a Global Positioning System receiver, Multipath Effect can cause a stationary receiver's output to indicate as if it were randomly jumping about or creeping. When the unit is moving the jumping or creeping may be hidden, but it still degrades the displayed accuracy of location and speed.

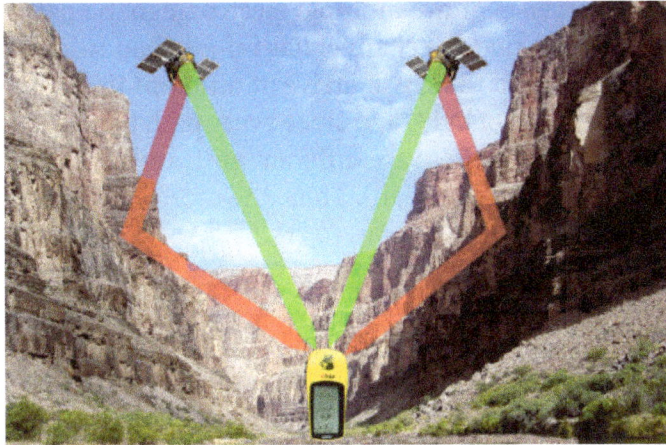

GPS error due to multipath

In Wired Media

Multipath propagation may also happen in wired media, especially where impedance mismatch causes signal reflection. A well-known example is power line communication.

High-speed power line communication systems usually employ multi-carrier modulations (such as OFDM or Wavelet OFDM) to avoid the intersymbol interference that multipath propagation would cause.

The ITU-T G.hn standard provides a way to create a high-speed (up to 1 Gigabit/s) local area network using existing home wiring (power lines, phone lines and coaxial cables). G.hn uses OFDM with a cyclic prefix to avoid ISI. Because multipath propagation behaves differently in each kind of wire, G.hn uses different OFDM parameters (OFDM symbol duration, Guard Interval duration) for each media.

Mathematical Modeling

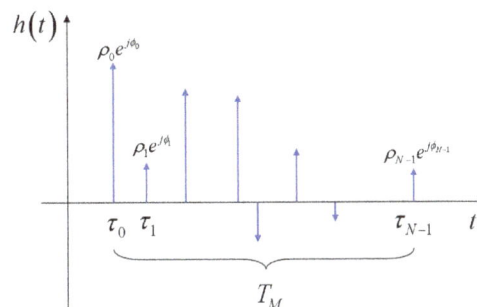

Mathematical model of the multipath impulse response.

The mathematical model of the multipath can be presented using the method of the impulse response used for studying linear systems.

Suppose you want to transmit a single, ideal Dirac pulse of electromagnetic power at time 0, i.e.

$$x(t) = \delta(t)$$

At the receiver, due to the presence of the multiple electromagnetic paths, more than one pulse will be received (we suppose here that the channel has infinite bandwidth, thus the pulse shape is not modified at all), and each one of them will arrive at different times. In fact, since the electromagnetic signals travel at the speed of light, and since every path has a geometrical length possibly different from that of the other ones, there are different air travelling times (consider that, in free space, the light takes 3 µs to cross a 1 km span). Thus, the received signal will be expressed by

$$y(t) = h(t) = \sum_{n=0}^{N-1} \rho_n e^{j\phi_n} \delta(t - \tau_n)$$

where N is the number of received impulses (equivalent to the number of electromagnetic paths, and possibly very large), τ_n is the time delay of the generic n^{th} impulse, and n^{th} represent the complex amplitude (i.e., magnitude and phase) of the generic received pulse. As a consequence, $y(t)$ also represents the impulse response function $h(t)$ of the equivalent multipath model.

More in general, in presence of time variation of the geometrical reflection conditions, this impulse response is time varying, and as such we have

$$\tau_n = \tau_n(t)$$

$$\rho_n = \rho_n(t)$$

$$\phi_n = \phi_n(t)$$

Very often, just one parameter is used to denote the severity of multipath conditions: it is called the multipath time, T_M, , and it is defined as the time delay existing between the first and the last received impulses

$$T_M = \tau_{N-1} - \tau_0$$

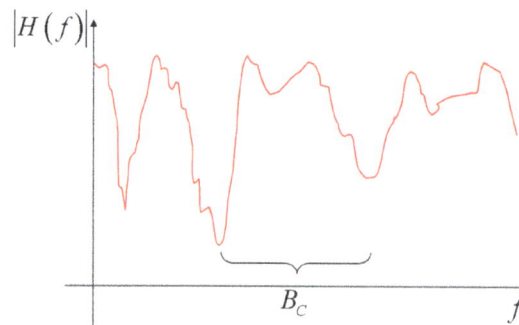

Mathematical model of the multipath channel transfer function.

In practical conditions and measurement, the multipath time is computed by considering as last impulse the first one which allows to receive a determined amount of the total transmitted power (scaled by the atmospheric and propagation losses), e.g. 99%.

Keeping our aim at linear, time invariant systems, we can also characterize the multipath phenomenon by the channel transfer function $H(f)$, which is defined as the continuous time Fourier transform of the impulse response $h(t)$

$$H(f) = \mathfrak{F}(h(t)) = \int_{-\infty}^{+\infty} h(t)e^{-j2\pi ft}\,dt = \sum_{n=0}^{N-1} \rho_n e^{j\phi_n} e^{-j2\pi f\tau_n}$$

where the last right-hand term of the previous equation is easily obtained by remembering that the Fourier transform of a Dirac pulse is a complex exponential function, an eigenfunction of every linear system.

The obtained channel transfer characteristic has a typical appearance of a sequence of peaks and valleys (also called *notches*); it can be shown that, on average, the distance (in Hz) between two consecutive valleys (or two consecutive peaks), is roughly inversely proportional to the multipath time. The so-called coherence bandwidth is thus defined as

$$B_C \approx \frac{1}{T_M}$$

For example, with a multipath time of 3 µs (corresponding to a 1 km of added on-air travel for the last received impulse), there is a coherence bandwidth of about 330 kHz.

Tropospheric Scatter

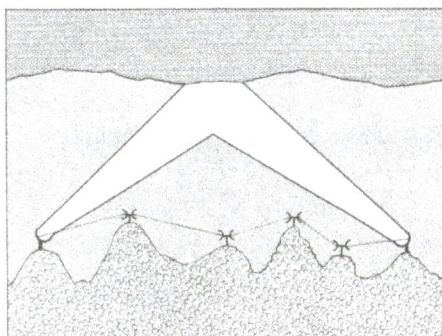

A tropospheric scatter system can bridge large distances while a microwave relay system *(below)* requires multiple relay stations due to its line of sight limitation.

Boswell Bay, Alaska White Alice Site, Tropospheric scatter antenna and feeder.

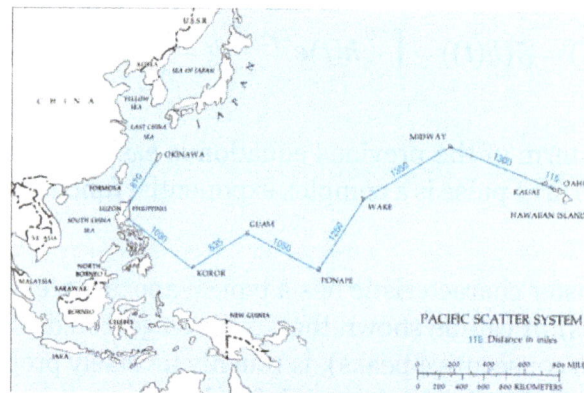

Pacific Scatter System

Tropospheric scatter (also known as troposcatter) is a method of communicating with microwave radio signals over considerable distances – often up to 300 km, and further depending on terrain and climate factors. This method of propagation uses the tropospheric scatter phenomenon, where radio waves at UHF and SHF frequencies are randomly scattered as they pass through the upper layers of the troposphere. Radio signals are transmitted in a narrow beam aimed just above the horizon in the direction of the receiver station. As the signals pass through the troposphere, some of the energy is scattered back toward the Earth, allowing the receiver station to pick up the signal.

Normally, signals in the microwave frequency range travel in straight lines, and so are limited to *line of sight* applications, in which the receiver can be 'seen' by the transmitter. So communication distances are limited by the visual horizon to around 30–40 miles. Troposcatter allows microwave communication beyond the horizon.

Because the troposphere is turbulent and has a high proportion of moisture the tropospheric scatter radio signals are refracted and consequently only a tiny proportion of the radio energy is collected by the receiving antennas. Frequencies of transmission around 2 GHz are best suited for tropospheric scatter systems as at this frequency the wavelength of the signal interacts well with the moist, turbulent areas of the troposphere, improving signal to noise ratios.

Overview

Historically, high gain dish or billboard antennas were required for tropospheric scatter systems as the propagation losses are very high; only about one billion-billionth (1×10^{-12}) of the transmit power is available at the receiver. Paths were established at distances over 1,000 km. They required antennas ranging from 9 meters to 36 meters and amplifiers ranging from 1 kW to 50 kW. These were analogue systems which were capable of transmitting a few voice channels.

Troposcatter systems have evolved over the years. With communication satellites used for long-distance communication links, current troposcatter systems are employed over shorter distances than previous systems, use smaller antennas and amplifiers, and have much higher bandwidth capabilities. Typical distances are between 50 km and 250 km, though greater distances can be achieved depending on the climate, terrain, and data rate required. Typical antenna sizes range from 1.2 meters to 12 meters while typical amplifier sizes range from 10W to 2 kW. Data rates over 20Mbit/s can be achieved with today's technology.

Tropospheric scatter is a fairly secure method of propagation as dish alignment is critical, making it extremely difficult to intercept the signals, especially if transmitted across open water, making them highly attractive to military users. Military systems have tended to be 'thin-line' tropo – so called because only a narrow bandwidth 'information' channel was carried on the tropo system; generally up to 32 analogue (4 kHz bandwidth) channels. Modern military systems are "Wideband" as they operate 4-16 Mbit/s digital data channels.

Civilian troposcatter systems, such as the British Telecom (BT) North Sea oil communications network, required higher capacity 'information' channels than were available using HF (high frequency – 3 to 30 MHz) radio signals, before satellite technology was available. The BT systems, based at Scousburgh in the Shetland Islands, Mormond Hill in Aberdeenshire and Row Brow near Scarborough, were capable of transmitting and receiving 156 analogue (4 kHz bandwidth) channels of data and telephony to / from North Sea oil production platforms, using frequency division multiplexing (FDMX) to combine the channels.

Because of the nature of the turbulence in the troposphere, quadruple diversity propagation paths were used to ensure 99.98% reliability of the service, equating to about 3 minutes of downtime due to propagation drop out per month. The quadruple space and polarisation diversity systems needed two separate dish antennae (spaced several metres apart) and two differently polarised feed horns – one using vertical polarisation, the other using horizontal polarisation. This ensured that at least one signal path was open at any one time. The signals from the four different paths were recombined in the receiver where a phase corrector removed the phase differences of each signal. Phase differences were caused by the different path lengths of each signal from transmitter to receiver. Once phase corrected, the four signals could be combined additively.

Tropospheric Scatter Communications Networks

The tropospheric scatter phenomenon has been used to build both civilian and military communication links in a number of parts of the world, including:-

ACE High

NATO in Europe.

BT (British Telecom)

United Kingdom - Shetland to Mormond Hill

Germany

Torfhaus-Berlin, Clenze-Berlin at Cold War times

Portugal Telecom

Portugal - Serra de Nogueira to Artzamendi

CNT (Canadian telecomms company)

Tsiigehtchic to Galena

Hay River - Port Radium - Lady Franklin Point

Cuba - Florida

> Guanabo to Florida City

AT&T Corporation

Project Offices

> Chatham, NC - Buckingham, VA - Charlottesville, VA - Leesburg, VA - Hagerstown, MD

Texas Towers - Air defence radars.

> Texas Tower 2
>
> Georges Shoal, in 56-foot (17 m) deep water, 110 miles (180 km) east of Cape Cod 41°44′N 67°47′W / 41.733°N 67.783°W / 41.733; -67.783 , linked to North Truro, MA.
>
> Texas Tower 3
>
> Nantucket Shoals, in 80-foot (24 m) water, 100 miles (160 km) south-east of Rhode Island 40°45′N 69°19′W / 40.75°N 69.317°W / 40.75; -69.317 , linked to Montauk AFB, Long Island, NY.
>
> Texas Tower 4
>
> Un-named Shoal (Unofficially: Old Shaky), in 185-foot (56 m) water, 84 miles (135 km) south-east of New York City 39°48′N 72°40′W (Destroyed, with 28 killed, during a storm on 15 January 1961), linked to Highlands, NJ mainland station.
>
> *Texas Tower 1*
>
> Cashes Ledge (Lat. 42° 53'N., Long. 68° 57'W., 36-foot depth), 100 miles east of New Hampshire, not built.
>
> *Texas Tower 5*
>
> Brown's Bank (Lat. 42° 47'N., Long. 65° 37'W., 84foot depth), 75 miles south of Nova Scotia, not built.

Mid Canada Line

> A series of five stations (070, 060, 050, 415, 410) in Ontario and Quebec around the lower Hudson Bay.

Pinetree Line, Pole Vault

> A series of fourteen stations providing communications for Eastern seaboard radar stations of the US/Canadian Pinetree line, running from N-31 Frobisher Bay, Baffin Island to St. John's, Newfoundland and Labrador.

White Alice

> A military and civil communications network with seventy-one stations stretching up the

western seaboard from Port Hardy, Vancouver island north to Barter Island (BAR) and east to Shemya, Alaska (SYA) in the Aleutian Islands.

DEW Training

A training facility for White Alice and the DEW line tropo-scatter network, between Pecatonica, Illinois to Streator, Illinois.

DEW Line

Several tropo-scatter networks providing communications for the extensive air-defence radar chain in the far north of Canada and the US.

NARS

NATO air-defence network stretching from RAF Fylingdales, via Mormond Hill, UK, Sornfelli (Faroe Islands), Höfn, Iceland to Keflavik DYE-5, Rockville.

ET-A, USAREUR

A US Army network from RAF Fylingdales to a network in Germany and a single station in France (Maison Fort).

486L, MEDCOM

A US Navy network covering the European coast of the Mediterranean Sea from San Pablo, Spain in the west to Adana AFB, Turkey in the East, with headquarters at Ringstead in Surrey, England.

Royal Air Force

Communications to British Forces Germany, running from Swingate in Kent to Lammersdorf in Germany.

BARS

A Warsaw Pact tropo-scatter network stretching from near Rostok in the DDR (Deutsches Demokratisches Republik), Czechoslovakia, Hungary, Poland, Byelorussia USSR, Ukraine USSR, Romania and Bulgaria.

SEVER

A Soviet network stretching right across the USSR.

India - USSR

A single section from Srinigar, Kashmir, India to Dangara, Tajikistan, USSR.

Indian Air Force

An air-defence network covering the Northern borders of India with at least 32 stations.

Peace Ruby, Spellout, Peace Net

An air-defence network set up by the United States in Iran pre-revolution. Spellout built a radar and comms network in the north of iran. Peace Ruby built another air-defence network in the south and Peace net integrated the two networks.

Bahrain - UAE

A tropo-scatter system linking Al Manamah, Bahrain to Dubai, United Arab Emirates.

RAFO

A tropo-scatter communications system providing military comms to the former SOAF - Sultan of Oman's Air Force, (now RAFO - Royal Air Force of Oman), across the Sultanate of Oman.

RSAF

A Royal Saudi Air Force tropo-scatter network linking major airbases and population centres in Saudi Arabia.

Yemen

A single system linking Sana'a with Sa'dah.

BACK PORCH and IWCS

Two networks run by the United States linking military bases in Thailand and South Vietnam.

Phil - Tai - Oki

A system linking the Philippines with Taiwan.

Japanese Troposcatter Networks

Two networks linking Japanese islands from North to South.

Tactical Troposcatter Communication systems

As well as the permanent networks detailed above, there have been many tactical transportable systems produced by several countries:-

Soviet / Russian Troposcatter Systems

MNIRTI R-423-1 Brig-1/R-423-2A Brig-2A/R-423-1KF

MNIRTI R-444 Eshelon / R-444-7,5 Eshelon D

MNIRTI R-420 Atlet-D

NIRTI R-417 Baget/R-417S Baget S

NPP Radiosvyaz R-412 A/B/F/S TORF

MNIRTI R-410/R-410-5,5/R-410-7,5 Atlet / Albatros

MNIRTI R-408/R-408M Baklan

Peoples Republic of China (PRoC), Peoples Liberation Army (PLA) Troposcatter Systems

CETC TS-504 Troposcatter Communication System

CETC TS-510/GS-510 Troposcatter Communication System

Western Troposcatter Systems

AN/TRC-97 Troposcatter Communication System

AN/TRC-170 Tropospheric Scatter Microwave Radio Terminal

AN/GRC-201 Troposcatter Communication System

The U.S. Army and Air Force use tactical tropospheric scatter systems developed by Raytheon for long haul communications. The systems come in two configurations, the original "heavy tropo", and a newer "light tropo" configuration exist. The systems provide four multiplexed group channels and trunk encryption, and 16 or 32 local analog phone extensions. The U.S. Marine Corps also uses the same device, albeit an older version.

US Army TRC-170 Tropo Scatter Microwave System

Line-of-sight Propagation

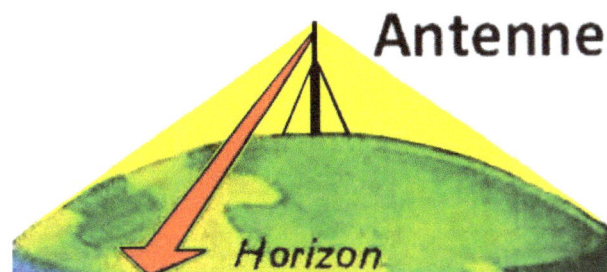

Line of sight propagation to an antenna

Line-of-sight propagation is a characteristic of electromagnetic radiation or acoustic wave propagation. Electromagnetic transmission includes light emissions traveling in a straight line. The rays or waves may be diffracted, refracted, reflected, or absorbed by atmosphere and obstructions with material and generally cannot travel over the horizon or behind obstacles.

At low frequency (below approximately 3 MHz), radio signals travel as ground waves, which follow the Earth's curvature due to diffraction with the layers of the atmosphere. This enables AM radio signals in low-noise environments to be received well after the transmitting antenna has dropped below the horizon. Additionally, frequencies between approximately 1 and 30 MHz can be reflected by the F1/F2 Layer, thus giving radio transmissions in this range a potentially global reach, again along multiple deflected straight lines. The effects of multiple diffraction or reflection lead to macroscopically "quasi-curved paths".

However, at frequencies above 30 MHz (VHF and higher) and in lower levels of the atmosphere, neither of these effects are significant. Thus, any obstruction between the transmitting antenna (transmitter) and the receiving antenna (receiver) will block the signal, just like the light that the eye may sense. Therefore, since the ability to visually see a transmitting antenna (disregarding the limitations of the eye's resolution) roughly corresponds to the ability to receive a radio signal from it, the propagation characteristicat these frequencies is called "line-of-sight". The farthest possible point of propagation is referred to as the "radio horizon".

In practice, the propagation characteristics of these radio waves vary substantially depending on the exact frequency and the strength of the transmitted signal (a function of both the transmitter and the antenna characteristics). Broadcast FM radio, at comparatively low frequencies of around 100 MHz, are less affected by the presence of buildings and forests.

Impairments to Line-of-sight Propagation

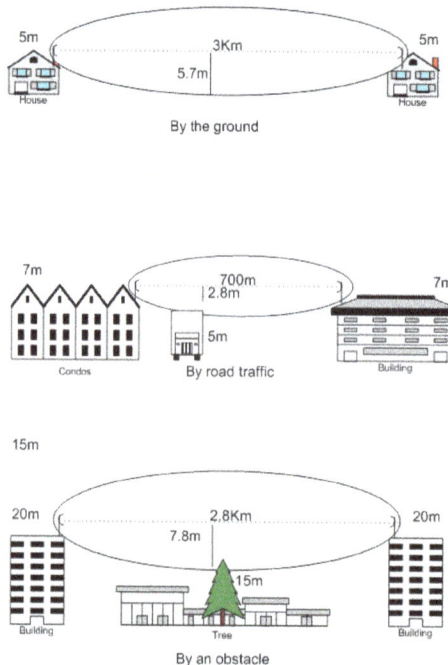

Objects within the Fresnel zone can disturb line of sight propagation even if they don't block the geometric line between antennas

Low-powered microwave transmitters can be foiled by tree branches, or even heavy rain or snow.

If a direct visual fix cannot be taken, it is important to take into account the curvature of the Earth when calculating line-of-sight from maps. Designs for microwave used to use 4/3 earth radius to compute clearances along the path.

The presence of objects not in the direct visual line of sight can interfere with radio transmission. This is caused by diffraction effects: for the best propagation, a volume known as the first Fresnel zone should be kept free of obstructions.

Reflected radiation from the ground plane also acts to cancel out the direct signal. This effect, combined with the free-space r^{-2} propagation loss to a r^{-4} propagation loss.This effect can be reduced by raising either or both antennas further from the ground: the reduction in loss achieved is known as *height gain*.

Mobile Telephones

Although the frequencies used by mobile phones (cell phones) are in the line-of-sight range, they still function in cities. This is made possible by a combination of the following effects:

- r–4 propagation over the rooftop landscape
- diffraction into the "street canyon" below
- multipath reflection along the street
- diffraction through windows, and attenuated passage through walls, into the building
- reflection, diffraction, and attenuated passage through internal walls, floors and ceilings within the building

The combination of all these effects makes the mobile phone propagation environment highly complex, with multipath effects and extensive Rayleigh fading. For mobile phone services, these problems are tackled using:

- rooftop or hilltop positioning of base stations
- many base stations (usually called "cell sites"). A phone can typically see at least three, and usually as many as six at any given time.
- "sectorized" antennas at the base stations. Instead of one antenna with omnidirectional coverage, the station may use as few as 3 (rural areas with few customers) or as many as 32 separate antennas, each covering a portion of the circular coverage. This allows the base station to use a directional antenna that is pointing at the user, which improves the signal to noise ratio. If the user moves (perhaps by walking or driving) from one antenna sector to another, the base station automatically selects the proper antenna.
- rapid handoff between base stations (roaming)
- the radio link used by the phones is a digital link with extensive error correction and detection in the digital protocol

- sufficient operation of mobile phone in tunnels when supported by split cable antennas

- local repeaters inside complex vehicles or buildings

A Faraday cage is composed of a conductor that completely surrounds an area on all sides, top, and bottom. Electromagnetic radiation is blocked where the wavelength is longer than any gaps. This means that windowless metal enclosures will completely block cell signs, such as elevator cabins, and parts of trains, cars, and ships. The same problem can affect signals in buildings with extensive steel reinforcement.

Line-of-sight Propagation as a Prerequisite for Radio Distance Measurements

The travel time of radio waves between transmitters and receivers can be measured disregarding the type of propagation. But, generally, travel time only then represents the distance between the transmitter and receiver when line of sight propagation is the basis for the measurement. This also applies to radar, Real Time Locating and lidar.

This rules: Travel time measurements for determining the distance between pairs of transmitters and receivers generally require line of sight propagation for proper results. Whereas the desire to have just any type of propagation to enable communication may suffice, this does never coincide with the requirement to have strictly line of sight at least temporarily as the means to obtain properly measured distances. However, the travel time measurement may always be biased by multi-path propagation, including line of sight propagation as well as non line of sight propagation in any random share. A qualified system for measuring the distance between transmitters and receivers must take this phenomenon into account. Thus filtering signals traveling along various paths makes the approach either operationally sound or just tediously irritating.

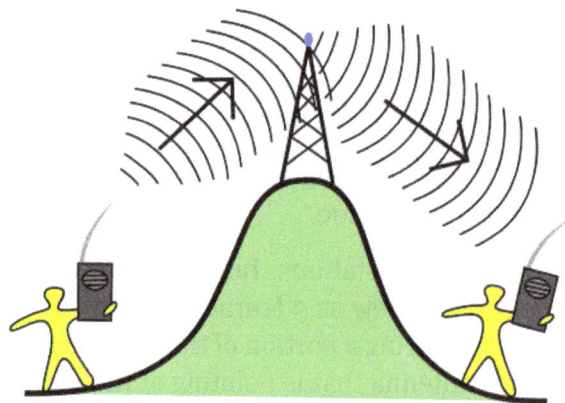

Two stations not in line-of-sight may be able to communicate through an intermediate radio repeater station.

Radio Horizon

The *radio horizon* is the locus of points at which direct rays from an antenna are tangential to the surface of the Earth. If the Earth was a perfect sphere and there was no atmosphere, the radio horizon would be a circle.

The radio horizon of the transmitting and receiving antennas can be added together to increase the effective communication range. Antenna heights above 1,000,000 feet (189 miles; 305 kilometres) will cover the entire hemisphere and not increase the radio horizon.

Radio wave propagation is affected by atmospheric conditions, ionospheric absorption, and the presence of obstructions, for example mountains or trees. Simple formulas that include the effect of the atmosphere give the range as:

$$\text{horizon}_{\text{miles}} \approx 1.23 \cdot \sqrt{\text{height}_{\text{feet}}} \;.$$

$$\text{horizon}_{\text{km}} \approx 3.57 \cdot \sqrt{\text{height}_{\text{metres}}}$$

The simple formulas give a best-case approximation of the maximum propagation distance, but are not sufficient to estimate the quality of service at any location.

Earth Bulge and Atmosphere Effect

Earth bulge is a term used in telecommunications. It refers to the circular segment of earth profile that blocks off long distance communications. Since the geometric line of sight passes at varying heights over the Earth, the propagating radio wave encounters slightly different propagation conditions over the path. The usual effect of the declining pressure of the atmosphere with height is to bend radio waves down towards the surface of the Earth, effectively increasing the Earth's radius, and the distance to the radio horizon, by a factor of around 4/3. This *k-factor* can change from its average value depending on weather.

Geometric Distance to Horizon

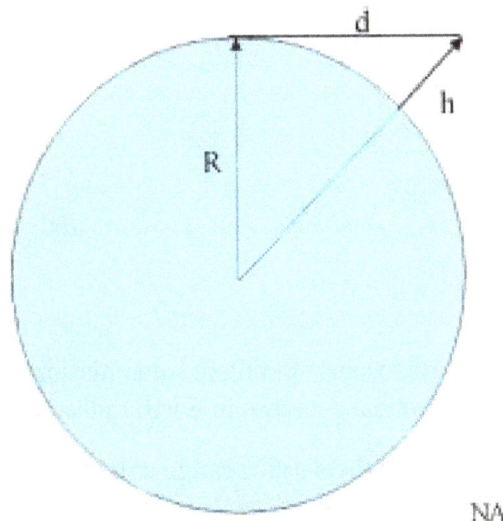

R is the radius of the Earth, h is the height of the transmitter (exaggerated), d is the line of sight distance

Assuming a perfect sphere with no terrain irregularity, the distance to the horizon from a high altitude transmitter (i.e., line of sight) can readily be calculated.

Let R be the radius of the Earth and h be the altitude of a telecommunication station. The line of sight distance d of this station is given by the Pythagorean theorem;

$$d^2 = (R+h)^2 - R^2 = 2 \cdot R \cdot h + h^2$$

Since the altitude of the station is much less than the radius of the Earth,

$$d \approx \sqrt{2 \cdot R \cdot h}$$

If the height is given in metres, and distance in kilometres,

$$d \approx 3.57 \cdot \sqrt{h}$$

If the height is given in feet, and the distance in miles,

$$d \approx 1.23 \cdot \sqrt{h}$$

The Actual Service Range

The above analysis does not consider the effect of atmosphere on the propagation path of RF signals. In fact, RF signals don't propagate in straight lines. Because of the refractive effects of atmospheric layers, the propagation paths are somewhat curved. Thus, the maximum service range of the station is not equal to the line of sight (geometric) distance. Usually, a factor k is used in the equation above

$$d \approx \sqrt{2 \cdot k \cdot R \cdot h}$$

k > 1 means geometrically reduced bulge and a longer service range. On the other hand, k < 1 means a shorter service range.

Under normal weather conditions, k is usually chosen to be 4/3. That means that the maximum service range increases by 15%.

$$d \approx 4.12 \cdot \sqrt{h}$$

for h in meters and d in kilometres; or

$$d \approx 1.41 \cdot \sqrt{h}$$

for h in feet and d in miles.

But in stormy weather, k may decrease to cause fading in transmission. (In extreme cases, k can be less than 1.) That is equivalent to a hypothetical decrease in Earth radius and an increase of Earth bulge.

ExampleIn normal weather conditions, the service range of a station at an altitude of 1500 m. with respect to receivers at sea level can be found as,

$$d \approx 4.12 \cdot \sqrt{1500} = 160 \text{ km.}$$

Two-way Radio

Several modern two-way hand-held radios compatible with the Project 25 digital radio standard
(Mobile and base station radios not shown)

A two-way radio is a radio that can both transmit and receive (a transceiver), unlike a broadcast receiver which only receives content. A two-way radio (transceiver) allows the operator to have a conversation with other similar radios operating on the same radio frequency (channel). Two-way radios are available in mobile, stationary base and hand-held portable configurations. Hand-held radios are often called walkie-talkies, handie-talkies, or just hand-helds.

Two-way radio systems usually operate in a half-duplex mode; that is, the operator can talk, or he can listen, but not at the same time. A push-to-talk or Press To Transmit button activates the transmitter; when it is released the receiver is active. A mobile phone or cellular telephone is an example of a two-way radio that both transmits and receives at the same time, ie in full-duplex mode. Full-duplex is generally achieved by the use of two different frequencies or by frequency-sharing methods to carry the two directions of the conversation simultaneously. Methods for mitigating the self interference caused by simultaneous same-frequency transmission and reception include using two antennas, or dynamic solid-state filters.

History

Installation of receivers and transmitters at the same fixed location allowed exchange of messages wirelessly. As early as 1907, two-way telegraphy traffic across the Atlantic Ocean was commercially available. By 1912 commercial and military ships carried both transmitters and receivers, allowing two-way communication in close to real-time with a ship that was out of sight of land.

The first truly mobile two-way radio was developed in Australia in 1923 by Senior Constable Frederick William Downie of the Victorian Police. The Victoria Police were the first in the world to use wireless communication in cars, putting an end to the inefficient status reports via public telephone boxes which had been used until that time. The first sets took up the entire back seat of the Lancia patrol cars.

As radio equipment became more powerful, compact, and easier to use, smaller vehicles had two-way radio communication equipment installed. Installation of radio equipment in aircraft allowed scouts to report back observations in real-time, not requiring the pilot to drop messages to troops on the ground below or to land and make a personal report.

In 1933, the Bayonne, New Jersey police department successfully operated a two-way system between a central fixed station and radio transceivers installed in police cars; this allowed rapidly directing police response in emergencies. During World War II walkie-talkie hand-held radio transceivers were extensively used by air and ground troops, both by the Allies and the Axis.

Early two-way schemes allowed only one station to transmit at a time while others listened, since all signals were on the same radio frequency – this was called "simplex" mode. Code and voice operations required a simple communication protocol to allow all stations to cooperate in using the single radio channel, so that one station's transmissions were not obscured by another's. By using receivers and transmitters tuned to different frequencies, and solving the problems introduced by operation of a receiver immediately next to a transmitter, simultaneous transmission and reception was possible at each end of a radio link, in so-called "full duplex" mode.

The first radio systems could not transmit voice. This required training of operators in use of Morse code. On a ship, the radio operating officers (sometimes shortened to "radio officers") typically had no other duties than handling radio messages. When voice transmission became possible, dedicated operators were no longer required and two-way use became more common. Today's two-way mobile radio equipment is nearly as simple to use as a household telephone, from the point of view of operating personnel, thereby making two-way communications a useful tool in a wide range of personal, commercial and military roles.

Types

Two-way radio systems can be classified in several ways depending on their attributes.

Conventional Versus Trunked

Conventional

Conventional radios operate on fixed RF channels. In the case of radios with multiple channels, they operate on one channel at a time. The proper channel is selected by a user. The user operates a channel selector (dial or buttons) on the radio control panel to pick the appropriate channel.

In multi-channel systems, channels are used for separate purposes. A channel may be reserved for a specific function or for a geographic area. In a functional channel system, one channel may allow City of Springfield road repair crews to talk to the City of Springfield's road maintenance office. A second channel may allow road repair crews to communicate with state highway department crews. In a geographic system, a taxi company may use one channel to communicate in the Boston, Massachusetts area and a second channel when taxis are in Providence, Rhode Island. In marine radio operations, one channel is used as an emergency and calling channel, so that stations may make contact then move to a separate working channel for continued communication.

Motorola uses the term *mode* to refer to channels on some conventional two-way radio models. In this use, a mode consists of a radio frequency channel and all channel-dependent options such as selective calling, channel scanning, power level, and more.

Scanning in Conventional Radios

Some conventional radios scan more than one channel. That is, the receiver searches more than one channel for a valid transmission. A valid transmission may be a radio channel with any signal or a combination of a radio channel with a specific CTCSS (or Selective calling) code.

There are a wide variety of scan configurations which vary from one system to another. Some radios have scan features that receive the primary selected channel at full volume and other channels in a scan list at reduced volume. This helps the user distinguish between the primary channel and others without looking at the radio control panel. An overview:

- A scanning feature can be defined and preset: when in scanning mode, a predetermined set of channels is scanned. Channels are not changeable by the radio user.

- Some radios allow an option for user-selected scan: this allows either lockout of pre-selected channels or adding channels to a scan list by the operator. The radio may revert to a default scan list each time it is powered off or may permanently store the most recent changes.

In professional radios, scan features are programmable and have many options. Scan features can affect system latency. If the radio has a twenty channel scan list and some channels have CTCSS, it can take several seconds to search the entire list. The radio must stop on each channel with a signal and check for a valid CTCSS before resuming scanning. This can cause missed messages.

For this reason, scan features are either not used or scan lists are intentionally kept short in emergency applications. Part of APCO Project 16 set standards for channel access times and delays caused by system overhead. Scan features can further increase these delays. One study said delays of longer than 0.4 seconds (400 milliseconds) in emergency services are not recommended. No delay from user push-to-talk until the user's voice is heard in the radio's speaker is an unattainable ideal.

Talk-back on Scan

Some conventional radios use, or have an option for, a talk-back-on-scan function. If the user transmits when the radio is in a scan mode, it may transmit on the last channel received instead of the selected channel. This may allow users of multi-channel radios to reply to the last message without looking at the radio to see which channel it was on. Without this feature, the user would have to use the channel selector to switch to the channel where the last message occurred. (This option can cause confusion and users must be trained to understand this feature.)

This is an incomplete list of some conventional radio types:

- Commercial and Public Safety Radio

- Marine VHF radio

- Family Radio Service (sometimes referred to by the abbreviation FRS)

- UNICOM

- Amateur Radio

Trunked

In a trunked radio system, the system logic automatically picks the *physical* radio frequency channel. There is a protocol that defines a relationship between the radios and the radio backbone which supports them. The protocol allows channel assignments to happen automatically.

Digital trunked systems may carry simultaneous conversations on one physical channel. In the case of a digital trunked radio system, the system also manages time slots on a single physical channel. The function of carrying simultaneous conversations over a single channel is called multiplexing.

Instead of channels, radios are related by groups which may be called, groups, talk groups, or divided into a hierarchy such as fleet and subfleet, or agency-fleet-subfleet. These can be thought of as virtual channels which appear and disappear as conversations occur.

Systems make arrangements for handshaking and connections between radios by one of these two methods:

- A computer assigns channels over a dedicated *control channel*. The control channel sends a continual data stream. All radios in the system monitor the data stream until commanded by the computer to join a conversation on an assigned channel.

- Electronics embedded in each radio communicate using a protocol of tones or data in order to establish a conversation, (scan-based).

If all physical channels are busy, some systems include a protocol to queue or stack pending requests until a channel becomes available.

Some trunked radios scan more than one talk group or agency-fleet-subfleet.

Visual clues a radio may be trunked include the 1) lack of a squelch knob or adjustment, 2) no *monitor* button or switch, and 3) a chirp (made famous by Nextel) showing the channel is available and ready at the moment the push-to-talk is pressed.

This is an incomplete list of some trunked radio types:

- TETRA

- Logic Trunked Radio (abbreviated LTR)

- SmartZone and SmartNet

- EDACS

Simplex Versus Duplex Channels

Simplex

Simplex channel systems use a single channel for transmit and receive. This is typical of aircraft

VHF AM, Citizens Band and marine radios. Simplex systems are often legacy systems that have existed since the 1930s. The architecture allows old radios to work with new ones in a single network. In the case of all ships worldwide or all aircraft worldwide, the large number of radios installed, (the *installed base*,) can take decades to upgrade. Simplex systems often use *open architectures* that allow any radio meeting basic standards to be compatible with the entire system.

- Advantage: as the simplest system configuration, there is reliability since only two radios are needed to establish communication between them, without any other infrastructure.

- Disadvantages: The simplex configuration offers communication over the shortest range or distance because mobile units must be in effective range of each other. The available channel bandwidth limits the number of simultaneous conversations, since "dead" air time cannot be easily used for additional communication.

Duplex

Duplex channel systems transmit and receive on different discrete channels. This defines systems where equipment cannot communicate without some infrastructure such as a repeater, base station or Talk-Through Base. Most common in the US is a repeater configuration where a base station is configured to simultaneously re-transmit the audio received from mobile units. This makes the mobiles, or hand-helds, able to communicate amongst one another anywhere within reception range of the base station or repeater. Typically the base or repeater station has a high antenna and high power, which allows much greater range, compared with a ground vehicle or hand-held transceiver.

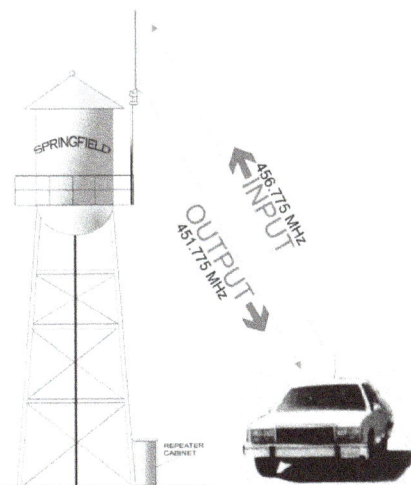

Duplex means two channels are used: one in each direction.

Duplex systems can be divided into two types. The term *half-duplex* refers to systems where use of a push-to-talk switch is required to communicate. *Full duplex* refers to systems like mobile telephones with a capability to simultaneously receive and transmit. Repeaters are by nature full duplex, most mobiles and almost all handhelds are half duplex.

- Advantage: duplex channels usually allow repeater operation which extends range (in most cases due to increased transmit power and improved aerial location / height) – especially where hand-held radios are in use.

- Disadvantage: If a radio cannot reach the repeater, it cannot communicate.

Hybrid Simplex/Duplex

Some systems use a mix of the two where radios use duplex as a default but can communicate simplex on the base station channel if out-of-range. In the US, the capability to talk simplex on a duplex channel with a repeater is sometimes called talk-around, direct, or car-to-car.

Push-to-Talk (PTT)

In two-way radios with headsets, a push-to-talk button may be included on a cord or wireless electronics box clipped to the user's clothing. In fire trucks or an ambulance a button may be present where the corded headset plugs into the radio wiring. Aircraft typically have corded headsets and a separate push-to-talk button on the control yoke or control stick. Dispatch consoles often have a hand-operated push-to-talk buttons along with a foot switch or pedal. If the dispatcher's hands are on a computer keyboard, the user can step on the foot pedal to transmit. Some systems have muting so the dispatcher can be on a telephone call and the caller cannot hear what is said over the radio. Their headset microphone will mute if they transmit. This relieves the dispatcher of explaining every radio message to a caller.

In some circumstances, voice-operated transmit (VOX) is used in place of a push-to-talk button. Possible uses are handicapped users who cannot push a button, Amateur radio operators, firefighters, crane operators, or others performing critical tasks where hands must be free but communication is still necessary.

Analog Versus Digital

One example of analog radios are AM aircraft radios used to communicate with control towers and air traffic controllers. Another is a Family Radio Service walkie talkie. Analog equipment is less complex than the simplest digital.

- Advantage: In high-quality equipment, better ability to communicate in cases where a received signal is weak or noisy.

- Disadvantage: Only one conversation at a time can occur on each channel.

Examples of digital communication technologies are all modern cellphones plus TETRA considered to be the best standard in digital radio and being the baseline infrastructure for whole of country networks, including manufacturers such as DAMM, Rohill, Cassidian, Sepura and others, APCO Project 25, a standard for digital public safety radios, and finally other systems such as Motorola's MotoTR-BO, Nextel's iDEN, Hytera's DMR, EMC's DMR, and NXDN implemented by Icom as IDAS and by Kenwood as NEXEDGE. Only NXDN and Mototrbo are proprietary DMR is an ETSI open standard.

- Advantage: More simultaneous talking paths are possible and information such as unit ID, status buttons, or text messages can be embedded into a single digital radio channel. The interoperability standard of TETRA means that any brand TETRA radio can work with any Brand TETRA infrastructure, not locking the user into expensive and proprietary systems.

- Disadvantage: Radios must be designed to the same, compatible standard, radios can become obsolete quickly (although this is mitigated by properly implemented interoperability standards such as those set down by ETSI for TETRA), cost more to purchase, and are more complicated.

Data Over Two-way Radio

In some cases, two-way radio is used to communicate analog or digital data. Systems can be simplex or duplex and may employ selective calling features such as CTCSS. In full-duplex systems, data can be sent real-time between two points. In simplex or half-duplex, data can be sent with a time lag between many points.

Some two-way digital systems carry both audio and data over a single data stream. Systems of this type include NXDN and APCO Project 25. Other more advanced systems under the TETRA standard are capable of joining time slots together to improve data bandwidth, allowing advanced data polling and telemetry applications over radio. The method of encoding and decoding the audio stream is called a codec, such as the AMBE or the ACELP family of codecs.

After market GPS tracking and mobile messaging devices can be interfaced with popular two-way radio models providing a range of features.

Analog

Analog systems may communicate a single condition, such as water level in a livestock tank. A transmitter at the tank site continually sends a signal with a constant audio tone. The tone would change in pitch to indicate the tank's water level. A meter at the remote end would vary, corresponding to the tone pitch, to indicate the amount of water present in the livestock tank. Similar methods can be used to telemeter any analog condition. This type of radio system serves a purpose equivalent to a four-to-twenty milliampere loop. In the US, mid-band 72–76 MHz or UHF 450–470 MHz interstitial channels are often used for these systems. Some systems multiplex telemetry of several analog conditions by limiting each to a separate range of tone pitches, for example.

Digital

Digital systems may communicate text messages from computer-aided dispatch (CAD). For example, a display in a tow truck may give a textual location for a call and any related details. The tow truck driver may press an *acknowledge* button, sending data in the opposite direction and flagging the call as received by the driver. They can be used for analog telemetry systems, such as the livestock tank levels, as described above. Another possibility is the lubricating oil pressure in a transit bus engine, or the current speed of the bus. Analog conditions are translated into data words. Some systems send radio paging messages which can either 1) beep a paging receiver, 2) send a numeric message, or 3) send a text message.

Digital systems typically use data rates in the 1,200–19,200 kilobit-per-second rates and may employ modulation schemes such as frequency shift keying, audio frequency shift keying, or quadrature phase shift keying to encode characters. Modern equipment have the same capabilities to carry data as are found in Internet Protocol. Working within the system's protocol constraints, virtually anything can be sent or received.

Engineered Versus Not Engineered

Engineered systems are designed to perform close to a specification or standard. They are designed as systems with all equipment matched to perform together. For example, a modern, local government two-way radio system in the US may be designed to provide 95% area coverage in an urban area. System designers use radio frequency models, terrain models, and signal propagation modeling software in an attempt to accurately estimate where radios will work within a defined geographic area. The models help designers choose equipment, equipment locations, antennas, and estimate how well signals will penetrate buildings. These models will be backed-up by drive testing and actual field signal level measurements. Designers adjust antenna patterns, add or move equipment sites, and design antenna networks in a way that will accomplish the intended level of performance.

Some systems are not engineered. *Legacy* systems are existing systems which were never designed to meet a system performance objective. They may have started with a base station and a group of mobile radios. Over a period of years, they have equipment added on in a building block style. Legacy systems may perform adequately even though they were not professionally designed as a coherent system. A user may purchase and locate a base station with an expectation that similar systems used in the past worked acceptably. A City Road Department may have a system that works acceptably, so the Parks Department may build a new similar system and find it equally usable. General Mobile Radio Service systems are not usually engineered.

Options, Duty Cycle, and Configuration

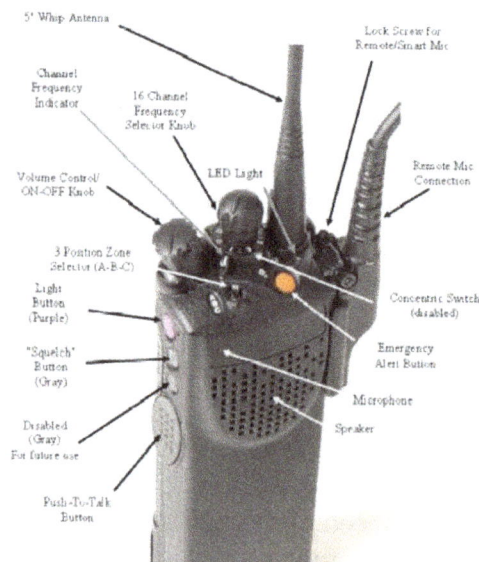

Example of control arrangement on a configured P25-capable hand-held radio.

1940s tube-type land mobile two way radios often had one channel and were carrier squelch. Because radios were costly and there were fewer radio users, it might be the case that no one else nearby used the same channel. A transmit and receive crystal had to be ordered for the desired channel frequency, then the radio had to be tuned or aligned to work on the channel. 12-volt mobile, tube-type radios drew several amperes on standby and tens-of-amperes on transmit. Equipment worked ideally when new. The performance of vacuum tubes gradually degraded over time. US regulations required an indicator lamp showing the transmitter had power applied and was ready to transmit and a second indicator, (usually red,) that showed the transmitter was on. In radios with options, wire jumpers and discrete components were used to select options. To change a setting, the technician soldered an option jumper wire then made any corresponding adjustments.

Many mobile and handhelds have a limited duty cycle. Duty Cycle is the ratio of listening time to transmit time and is generally dependent on how well the transmitter can shed the heat from the heat sink on the rear of the radio. A 10% duty cycle (common on handhelds) translates to 10 seconds of transmit time to 90 seconds of receive time. Some mobile and base equipment is specified at different power levels – for example 100% duty cycle at 25 watts and 15% at 40 watts.

The trend is toward increasing complexity. Modern handheld and mobile radios can have capacities as high as 255 channels. Most are synthesized: the internal electronics in modern radios operate over a range of frequencies with no tuning adjustments. High-end models may have several hundred optional settings and require a computer and software to configure. Sometimes, controls on the radio are referred to as programmable. By changing configuration settings, a system designer could choose to set up a button on the radio's control panel to function as:

- turn scan on or off,

- alert another mobile radio, (selective calling),

- turn on an outside speaker, or

- select repeater locations.

In most modern radios these settings are done with specialized software (provided by the manufacturer) and a connection to a laptop computer.

Microprocessor-based radios can draw less than 0.2 amperes on standby and up to tens-of-amperes on high-powered, 100 watt transmitters.

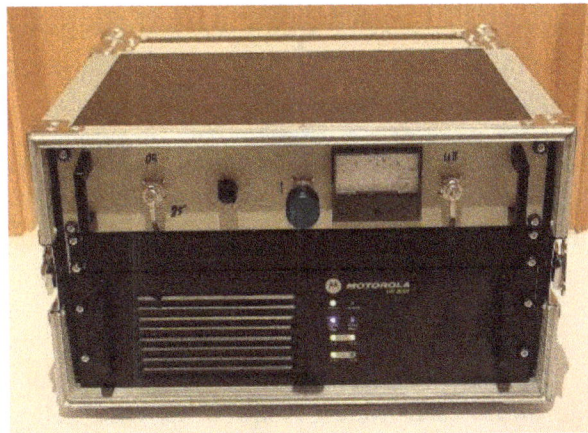

Motorola MOTOTRBO Repeater DR3000 with duplexer mounted in Flightcase, 100% Duty cycle up to 40 W output

Base stations, repeaters, and high-quality mobile radios often have specifications that include a duty cycle. A repeater should always be *continuous duty*. This means the radio is designed to transmit in a continuous broadcast without transmitter overheating and resulting failure. Handhelds are intermittent duty, mobile radios and base station radios are available in normal or continuous duty configurations. Continuous duty is preferred in mobile emergency equipment because any one of an entire fleet of ambulances, for example, could be pressed into service as command post at a major incident. Unfortunately budgets frequently get in the way and intermittent duty radios are purchased.

Time delay is always associated with radio systems, but it is apparent in spacecraft communications. NASA regularly communicates with exploratory spacecraft where a round-trip message time is measured in hours (like out past Jupiter). For Apollo program and Space Shuttle, Quindar tones were used for transmit PTT control.

Life of Equipment

Though the general life term for the two way radio is 5 to 7 years and 1 to 2 years for its accessories but still the usage, atmosphere and environment plays a major role to decide its life term (radios are often deployed in harsh environments where more fragile communication equipment such as phones and tablets may fail). There are so many speculations on the life term of two way radios and their accessories i.e. batteries, chargers, head set etc.

In government systems, equipment may be replaced based on budgeting rather than any plan or expected service life. Funding in government agencies may be cyclical or sporadic. Managers may replace computing systems, vehicles, or budget computer and vehicle support costs while ignoring two-way radio equipment. Equipment may remain in use even though maintenance costs are unreasonable when viewed from an efficiency standpoint.

Different system elements will have differing service lifetimes. These may be affected by who uses the equipment. An individual contacted at one county government agency claimed equipment used by 24-hour services wears out much faster than equipment used by those who work in positions staffed eight hours a day.

One document says "seven years" is beyond the expected lifetime of walkie-talkies in police service. Batteries are cited as needing replacement more often. Twelve-year-old dispatch consoles mentioned in the same document were identified as usable. These were compared to problematic 21-year-old consoles used elsewhere in the same system.

Another source says system backbone equipment like consoles and base stations are expected to have a fifteen-year life. Mobile radios are expected to last ten years. Walkie talkies typically last eight. In a State of California document, the Department of General Services reports expected service life for a communications console used in the Department of Forestry and Fire Protection is 10 years.

Two-way Radio Frequencies

Two-way radios can operate on many different frequencies, and these frequencies are assigned differently in different countries. Typically channelized operations are used, so that operators need not tune equipment to a particular frequency but instead can use one or more pre-selected frequencies, easily chosen by a dial, a pushbutton or other means. For example, in the United States, there is a block of 5 channels (pre-selected radio frequencies) are allocated to the Multiple Use Radio System. A different block of 22 channels are assigned, collectively, to the General Mobile Radio Service and Family Radio Service. The Citizens Radio Service (""CB"") has 40 channels.

In an analog, conventional system, (the simplest type of system) a frequency or channel serves as a physical medium or link carrying communicated information. The performance of a radio system is partly dependent on the characteristics of frequency band used. The selection of a frequency for a two-way radio system is affected, in part, by:

- government licensing and regulations.

- local congestion or availability of frequencies.

- terrain, since radio signals travel differently in forests and urban viewsheds.

- the presence of noise, interference, or intermodulation.

- sky wave interference below 50–60 MHz and tropospheric bending at VHF.

- in the US, some frequencies require approval of a frequency coordination committee.

A channel number is just a shorthand notation for a frequency. It is, for instance, easier to remember "Channel 1" than to remember "26.965 MHz" (US CB Channel 1) or "462.5625 MHz" (FRS/GMRS channel 1), or "156.05 MHz" (Marine channel 1). It is necessary to identify which radio service is under discussion when specifying a frequency by its channel number. Organizations, such as electric power utilities or police departments, may have several assigned frequencies in use with arbitrarily assigned channel numbers. For example, one police department's "Channel 1" might be known to another department as "Channel 3" or may not even be available. Public service agencies have an interest in maintaining some common frequencies for inter-area or inter-service coordination in emergencies (modern term: *interoperability*).

Each country allocates radio frequencies to different two-way services, in accordance with international agreements. In the United States some examples of two-way services are: Citizen's Band, FRS, GMRS, MURS, and BRS.

Amateur radio operators nearly always use frequencies rather than channel numbers, since there is no regulatory or operating requirement for fixed channels in this context. Even amateur radio equipment will have "memory" features to allow rapidly setting the transmitter and receiver to favorite frequencies.

UHF Versus VHF

The most common two-way radio systems operate in the VHF and UHF parts of the radio spectrum. Because this part of the spectrum is heavily used for broadcasting and multiple competing uses, spectrum management has become an important activity of governments to regulate radio users in the interests of efficient and non-interfering use of radio. Both bands are widely applied for different users.

UHF has a shorter wavelength which makes it easier for the signal to find its way through smaller wall openings to the inside of a building. The longer wavelength of VHF means it can transmit further under normal conditions. For most applications, lower radio frequencies are better for longer range and through vegetation. A broadcasting TV station illustrates this. A typical VHF TV station operates at about 100,000 watts and has a coverage radius range of about 60 miles. A UHF TV station with a 60-mile coverage radius requires transmitting at 3,000,000 watts. Another factor with higher frequencies (UHF) is that smaller sized objects will absorb or reflect the energy more which causes range loss and/or multipath reflections which can weaken a signal by causing an "Out of Time/Out of Phase" signal to reach the antenna of the receiver (this is what caused the "Ghost" image on old over the air television).

If an application requires working mostly outdoors, a VHF radio is probably the best choice, especially if a base station radio indoors is used and an external antenna is added. The higher the antenna is placed, the further the radio can transmit and receive.

If the radios are used mainly inside buildings, then UHF is likely the best solution since its shorter wavelength travels through small openings in the building better. There are also repeaters that can be installed that can relay any frequencies signal (VHF or UHF) to increase the communication distance.

There are more available channels with UHF. Since the range of UHF is also not as far as VHF under most conditions, there is less chance of distant radios interfering with the signal. UHF is less affected than VHF by manmade electrical noise. So as you see, radio technology is very dynamic and you must make the choice of what to use based on your individual situation.

Range

The useful direct range of a two-way radio system depends on radio propagation conditions, which are a function of frequency, antenna height and characteristics, atmospheric noise, reflection and refraction within the atmosphere, transmitter power and receiver sensitivity, and required signal-to-noise ratio for the chosen modulation method. An engineered two-way radio system will calculate the coverage of any given base station with an estimate of the reliability of the communication at that range. Two-way systems operating in the VHF and UHF bands, where many land mobile systems operate, rely on line-of-sight propagation for the reliable coverage area. The "shadowing" effect of tall buildings may block reception in areas within the line-of-sight range which can be achieved in open countryside free of obstructions. The approximate line-of-sight distance to the radio horizon can be estimated from : horizon in kilometers = 3.569 times the square root of the antenna height in meters.

There are other factors that affect the range of a two-way radio such as weather, exact frequency used, and obstructions.

Other Two-way radio Devices

Not all two way radios are hand-held devices. The same technology that is used in two way radios can be placed in other radio forms. An example of this is a wireless callbox. A wireless callbox is a device that can be used for voice communication at security gates and doors. Not only can they be used to talk to people at these entry points, personnel can remotely unlock the door so the visitor can enter. There are also customer service callboxes that can be placed around a business that a customer can use to summon help from a two-way radio equipped store employee.

Another use of two-way radio technology is for a wireless PA system. A wireless PA is essentially a one-way two way radio that enables broadcasting messages from handheld two-way radios or base station intercoms.

Two-way radio rental businessAs two-way radios became the leading method of two-way communication, industries like movie and television production companies, security companies, event companies, sporting events and others needed to find a solution to use two way radios that was cost-effective and economically smart. Instead of buying two-way radios these companies began renting two-way radios short term and long term. The two-way radio rentals is a significant and important component of two-way radio businesses. Many have become reluctant to buy two way radios because of the duration of their event or the necessity to save money. Renting two way radios has brought comfort to customers in renting two way radios because the price and non commitment to owning such two way communication devices. Customers can rent anything from two-way radios to two-way radio equipment like speaker microphones or repeaters.

Skip Zone

A skip zone, also called a silent zone or zone of silence, is a region where a radio transmission can not be received. The zone is located between regions both closer and farther from the transmitter where reception is possible.

When using medium to high frequency radio telecommunication, there are radio waves which travel both parallel to the ground, and towards the ionosphere, referred to as a ground wave and sky wave, respectively. A skip zone is an annular region between the farthest points at which the ground wave can be received and the nearest point at which the refracted sky waves can be received. Within this region, no signal can be received because, due to the conditions of the local ionosphere, the relevant sky waves are not reflected but penetrate the ionosphere.

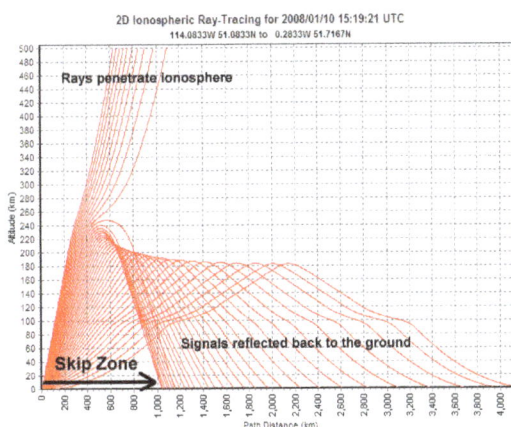

Formation of a skip-zone using Proplab-Pro 3.

The skip zone is a natural phenomenon that cannot be influenced by technical means. Its width depends on the height and shape of the ionosphere and, particularly, on the local ionospheric maximum electron density characterized by critical frequency f_oF_2. It varies mainly with this parameter, being larger for low f_oF_2. With a fixed working frequency it is large by night and may even disappear by day. Transmitting at night is most effective for long distance communication but the skip zone becomes significantly larger. Very high frequency waves and higher normally travel through the ionosphere wherefore communication via skywave is exceptional. A highly ionized Es-Layer that occasionally may appear in Summer may produce such an example.

Another method of decreasing the skip zone is by decreasing the frequency of the radio waves. Decreasing the frequency is akin to increasing the ionospheric width. A point is eventually reached when decreasing the frequency results in a zero distance skip zone. In other words, a frequency exists for which vertically incident radio waves will always be refracted back to the Earth. This frequency is equivalent to the ionospheric plasma frequency and is also known as the ionospheric critical frequency, or f_oF_2.

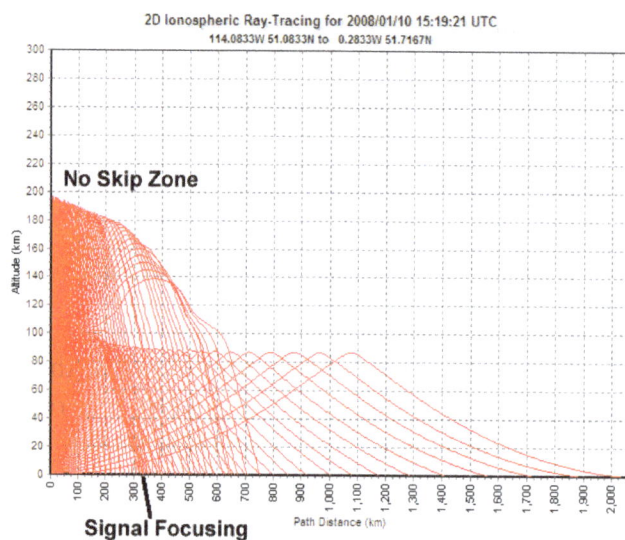

If the radio wave frequency is decreased, a point is reached where all waves (even vertically incident waves) are reflected back to the Earth.

Skip zone is the subject of a film 'SKIPZONE' made in 1992 by UK artist, Peter Lee-Jones. It refers to areas in Scottish Highlands where it is difficult to obtain radio and TV reception.

References

- Ulrich L. Rohde, Jerry Whitaker "Communications Receivers, Third Edition" McGraw Hill, New York, NY, 2001, ISBN 0-07-136121-9.

- Goldsmith, Andrea (8 Aug 2005). Wireless Communications. Cambridge University Press. ISBN 9780521837163. Retrieved 20 April 2016.

- Haldane, Robert. (1995) The People's Force, A history of the Victoria Police. Melbourne University Press. ISBN 0-522-84674-2, 1995

- Duarte, Melissa; Sabharwal, Ashutosh (2010). "Full-Duplex Wireless Communications Using Off-The-Shelf Radios: Feasibility and First Results" (PDF). WARP Project. Retrieved 20 April 2016.

Permissions

All chapters in this book are published with permission under the Creative Commons Attribution Share Alike License or equivalent. Every chapter published in this book has been scrutinized by our experts. Their significance has been extensively debated. The topics covered herein carry significant information for a comprehensive understanding. They may even be implemented as practical applications or may be referred to as a beginning point for further studies.

We would like to thank the editorial team for lending their expertise to make the book truly unique. They have played a crucial role in the development of this book. Without their invaluable contributions this book wouldn't have been possible. They have made vital efforts to compile up to date information on the varied aspects of this subject to make this book a valuable addition to the collection of many professionals and students.

This book was conceptualized with the vision of imparting up-to-date and integrated information in this field. To ensure the same, a matchless editorial board was set up. Every individual on the board went through rigorous rounds of assessment to prove their worth. After which they invested a large part of their time researching and compiling the most relevant data for our readers.

The editorial board has been involved in producing this book since its inception. They have spent rigorous hours researching and exploring the diverse topics which have resulted in the successful publishing of this book. They have passed on their knowledge of decades through this book. To expedite this challenging task, the publisher supported the team at every step. A small team of assistant editors was also appointed to further simplify the editing procedure and attain best results for the readers.

Apart from the editorial board, the designing team has also invested a significant amount of their time in understanding the subject and creating the most relevant covers. They scrutinized every image to scout for the most suitable representation of the subject and create an appropriate cover for the book.

The publishing team has been an ardent support to the editorial, designing and production team. Their endless efforts to recruit the best for this project, has resulted in the accomplishment of this book. They are a veteran in the field of academics and their pool of knowledge is as vast as their experience in printing. Their expertise and guidance has proved useful at every step. Their uncompromising quality standards have made this book an exceptional effort. Their encouragement from time to time has been an inspiration for everyone.

The publisher and the editorial board hope that this book will prove to be a valuable piece of knowledge for students, practitioners and scholars across the globe.

Index

www.ingramcontent.com/pod-product-compliance
Lightning Source LLC
Chambersburg PA
CBHW082058190326
41458CB00010B/3522